The Periodic Table

1								
1 1 H hydrogen 1.01	**2**							
2 3 Li lithium 6.94	4 Be beryllium 9.01							
3 11 Na sodium 22.99	12 Mg magnesium 24.31	**3**	**4**	**5**	**6**	**7**	**8**	**9**
4 19 K potassium 39.11	20 Ca calcium 40.08	21 Sc scandium 44.96	22 Ti titanium 47.88	23 V vanadium 50.94	24 Cr chromium 52.01	25 Mn manganese 54.94	26 Fe iron 55.85	27 Co cobalt 58.93
5 37 Rb rubidium 85.47	38 Sr strontium 87.62	39 Y yttrium 88.91	40 Zr zirconium 91.22	41 Nb niobium 92.91	42 Mo molybdenum 95.94	43 Tc technetium [99]	44 Ru ruthenium 101.07	45 Rh rhodium 102.91
6 55 Cs cesium 140.12	56 Ba barium 140.91	see below	72 Hf hafnium 178.49	73 Ta tantalum 180.95	74 W tungsten 183.85	75 Re rhenium 186.21	76 Os osmium 190.21	77 Ir iridium 192.22
7 87 Fr francium [223]	88 Ra radium [226]		104 Rf rutherfordium [267]	105 Db dubnium [268]	106 Sg seaborgium [271]	107 Bh bohrium [272]	108 Hs hassium [270]	109 Mt meitnerium [276]

LEGEND:
- ——–> Atomic Number
- ——–> Symbol
- ——–> Name
- ——–> Atomic Mass

Metals | Semimetals | Nonmetals

57 La lanthanum 138.91	58 Ce cerium 140.12	59 Pr praseodymium 140.91	60 Nd neodymium 144.24	61 Pm promethium [147]	62 Sm samarium 150.36	63 Eu europium 151.97
89 Ac actinium [227]	90 Th thorium [232]	91 Pa protactinium [231]	92 U uranium [238]	93 Np neptunium [237]	94 Pu plutonium [244]	95 Am americium [243]

of the Elements

			13	14	15	16	17	18
								2 He helium 4.00
	Solid		5 B boron 10.81	6 C carbon 12.01	7 N nitrogen 14.01	8 O oxygen 16.00	9 F fluorine 19.00	10 Ne neon 20.18
	Liquid		13 Al aluminum 26.98	14 Si silicon 28.09	15 P phosphorus 30.97	16 S sulfur 32.07	17 Cl chlorine 35.45	18 Ar argon 39.95
10	11	12						
28 Ni nickel 58.69	29 Cu copper 63.55	30 Zn zinc 65.39	31 Ga gallium 69.72	32 Ge germanium 72.61	33 As arsenic 74.92	34 Se selenium 78.96	35 Br bromine 79.91	36 Kr krypton 83.81
46 Pd palladium 106.42	47 Ag silver 107.87	48 Cd cadmium 112.41	49 In indium 114.82	50 Sn tin 118.71	51 Sb antimony 121.75	52 Te tellurium 127.61	53 I iodine 126.91	54 Xe xenon 131.29
78 Pt platinum 195.08	79 Au gold 196.97	80 Hg mercury 200.59	81 Tl thallium 204.38	82 Pb lead 207.21	83 Bi bismuth 208.98	84 Po polonium [209]	85 At astatine [210]	86 Rn radon [222]
110 Ds damstadtium [281]	111 Rg roentgenium [280]	112 Cn copernicium [285]	113 ?	114 ?	115 ?	116 ?	117 ?	118 ?

64 Gd gadolinium 157.25	65 Tb terbium 158.93	66 Dy dysprosium 162.51	67 Ho holmium 164.93	68 Er erbium 167.26	69 Tm thulium 168.93	70 Yb ytterbium 173.04	71 Lu lutetium 174.97
96 Cm curium [247]	97 Bk berkelium [247]	98 Cf californium [251]	99 Es einsteinium [252]	100 Fm fermium [257]	101 Md mendelevium [258]	102 No nobelium [259]	103 Lr lawrencium [260]

CONDENSED CHEMISTRY

Art Ingle
Maricopa County Community College District

Linus
Publications, Inc.

Published by Linus Publications, Inc.
Deer Park, NY 11729

Cover illustration by Steve Failla

Copyright © 2011 by Linus Publications
All Rights Reserved.

ISBN: 1-60797-223-9

No part of this publication may be reproduced, stored in a retrieval system, or transmitted, in any form or by any means, electronic, mechanical, photocopying, recording, or otherwise, without the prior permission of the publisher.

Printed in the United States of America.

Print Numbers   5   4   3   2   1

Table of Contents

Chapter 1
What is Matter?... 01

Chapter 2
What does an Atom look like?.. 11

Chapter 3
What is the Periodic Table?.. 23

Chapter 4
What is a Chemical Bond?.. 35

Chapter 5
What are Different Compounds called?... 49

Chapter 6
What are Chemical Equations?.. 61

Chapter 7
What is a Chemical Reaction?.. 69

Chapter 8
What is a Mole?... 81

Chapter 9
How many Moles?... 95

Chapter 10
What is a Gas? A Liquid? A Solid?... 107

Chapter 11
What are Solutions?.. 123

Chapter 12
What is an Acid? A Base?... 137

Chapter 13
What is Chemical Equilibrium?.. 151

Chapter 14
What are Oxidation Numbers?... 165

Bibliography 175

Foreword to the Instructor

My students always say "Tell me only what I need to know to pass the test". **Condensed Chemistry** is an attempt to do just that. It tries to present only the Need-to-Know information required to pass an introduction to chemistry course. All the other information I call Nice-to-Know, because that is just what it is. It consists of the history of chemistry, obscure terms and theories, explanations of archaic practices, and the like that the students are not going to remember anyway. Although this book does not waste the student's time with unnecessary details, at the same time it makes it essential that they carefully read and thoroughly understand every word, because it is ALL Need-to-Know information. At the start of the course please tell your students that they will need to read each chapter, or at least parts of it, twice and sometimes even three times.

Something else I hear from students is "Say it in words I can understand". **Condensed Chemistry** attempts to do that too. You may encounter overly simplified definitions, slang words, poor grammar, and even some dangling prepositions. Know that I do it on purpose. Over the years I have found that if you talk the way they talk, they understand it better and do better on tests. With my apologies, I leave it to their English instructor to try to teach them proper grammar and diction while I concentrate on teaching them chemistry.

An additional benefit of skipping all the Nice-to-Know information and using overly simplified definitions is that the chapters and the entire book are much shorter. That not only lowers the price of the book, but students are more likely to read a chapter 10 pages long than a chapter 40 pages long.

You will also notice that I skipped the scientific method, the metric system, taking measurements, and the like. I feel those subjects more properly belong in the laboratory part of the course, and I leave it to the lab instructor to take a few minutes at the start of the first few labs to present those topics.

Lastly, over the years I have found ways to present some concepts such that they seem to make better sense to the students. It gives me satisfaction to hear them say "I find the book very confusing, but I understand it when you explain it". My experience is they are not just complementing me to get a higher grade. They really do better on tests. I have tried to include some of my approaches in this book, and I hope that you too find that some concepts do make better sense to the students when presented in those ways.

Please note that this book is not an advanced or scholarly text in chemistry. It is simply an introduction to chemistry for non-chemistry majors. Anyone interested in advanced chemistry is advised to find another text book.

Your feedback and suggestions on this book will be greatly appreciated. You can write me at Chandler-Gilbert Community College in Chandler, AZ.

Art Ingle

Acknowledgements

Foundations of college chemistry by Hein, Morris. Reproduced with permission of JOHN WILEY & SONS INC. - BOOKS in the format Copy via Copyright Clearance Center.

Major Concepts and Key Skills taught in this Book

1. What is Matter?

- Characteristics of matter
- Light is not matter
- Physical forms of matter
- Physical versus chemical changes
- Heterogeneous mixtures
- Homogenous mixtures
- Mixtures and compounds
- Elements and compounds
- Chemical formulas
- Chemical reactions
- Branches of Chemistry

2. What does an Atom look like?

- Protons, electrons, neutrons
- The atomic nucleus
- Orbits and orbitals
- Energy levels of orbitals
- Electron excitations
- Photon emission
- Atomic spectra
- Electron configuration of atoms
- Write the electron configuration of an atom
- Electron configuration of ions
- Write the electron configuration of an ion
- Atomic notation
- Isotopes
- Two myths about atoms

3. What is the Periodic Table?

- Information about the elements
- General layout of the Periodic Table
- Groups and series
- Collective names for groups
- Collective names for series
- Information about the groups
- General trends in the Periodic Table
- Atomic diameter
- Metallic character
- Electro-negativity
- General trends down groups
- Predict the atomic mass of an element
- Electron configurations and the Periodic Table
- The Periodic Law

4. What is a Chemical Bond?

- Chemical bonds
- Chemical reactions
- Covalent bonds
- Polar covalent bonds
- Ionic bonds
- Valence electrons
- Determine the valence electrons of an element
- Electron dot formulas
- Write the electron dot formula of an element
- Lewis dot structures
- Write the Lewis dot structure of a compound
- Structural formulas
- Write the structural formula of a compound
- Polyatomic ions
- Valence shell electron pair geometry
- Determine the electron pair geometry of a compound
- Molecular shapes
- Determine the molecular shape of a compound

5. What are Different Compounds called?

- Inorganic compounds
- Organic compounds
- Naming binary molecular compounds
- Using the Greek number prefix system
- Name a binary molecular compound
- What about ternary molecular compounds
- Naming binary ionic compounds
- Using the Roman numeral (Stock) system
- Name a binary ionic compound
- Naming ternary ionic compounds
- Using –ate and –ite suffixes
- Using per- and hypo- prefixes
- Name a ternary ionic compound
- Naming binary acids
- Name a binary acid
- Naming ternary acids
- Using –ic and –ous suffixes
- Name a ternary acid
- Systematic names

6. What are Chemical Equations?

- Format of chemical equations
- Symbols in chemical equations
- Reactants and products
- How to write chemical equations
- Write a chemical equation
- How to balance chemical equations
- Balance a chemical equation
- Advanced chemical equations

7. What is a Chemical Reaction?

- Four signs of a chemical reaction
- Composition reactions
- Decomposition reactions
- Single replacement reactions
- The activity series
- Use the activity series
- Double replacement reactions
- The solubility table
- Use the solubility table
- Neutralization reactions

8. What is a Mole?

- The thing called a mole
- Conversion factors
- Use conversion factors
- Molar (Avogadro's) number
- Calculate moles from particles
- Molar volume of a gas
- Calculate moles from liters of a gas
- Molar mass of an element
- Calculate moles from grams
- Molar mass of a compound
- Determine molar mass of a compound
- Multiple conversions
- Calculate with multiple conversions
- Calculate the mass of a molecule
- Calculate the density of a gas
- Composition percents
- Calculate composition percents of a compound
- Hydrate complexes
- Calculate hydrate composition percents
- Empirical formulas
- Calculate an empirical formula
- Molecular formulas
- Calculate a molecular formula

9. How many Moles?

- Reactions at a mole level
- Mole-to-mole conversions
- Do mole-to-mole conversions
- Conversions with grams, liters, or particles
- Do mole-to-mole conversion from grams to liters
- Limiting and excess reactants
- Determine limiting and excess reactants
- Percent yield
- Calculate percent yield

10. What is a Gas? A Liquid? A Solid?

- General characteristics of gases
- Gases mix homogeneously
- The ideal gas law
- Use the ideal gas law
- Atmospheric pressure
- General characteristics of liquids
- Intramolecular and intermolecular bonds
- Dispersion bonds
- Dipole bonds
- Hydrogen bonds
- Surface tension of liquids
- Boiling point of liquids
- Vapor pressure of liquids
- Viscosity of liquids
- General characteristics of solids
- Crystalline structure
- Ionic crystals
- Molecular crystals
- Metallic crystals
- Melting point of solids
- Changes from a solid to a liquid to a gas
- Calculate the heat in changing steam to ice

XI

11. What are Solutions?

- Solutes and solvents
- Mass percent
- Calculate the mass percent of a solution
- Molarity
- Calculate the molarity of a solution
- Standard solutions
- Solubility
- Effect of temperature on solubility
- Find the effect of temperature on solubility
- Effect of pressure on solubility
- Calculate the effect of pressure on solubility
- Unsaturated and supersaturated solutions
- The process of dissolving
- Solvent cages and solvent blankets
- Polar and nonpolar compounds
- Miscible and immiscible liquids
- Factors affecting the rate of dissolving
- Dilutions of acids and bases
- Calculate the water needed to dilute an acid

12. What is an Acid? A Base?

- Arrhenius acids and bases
- Bronsted-Lowry acids and bases
- Major differences between them
- Measuring strength of an acid or base
- The pH scale
- Using four types of pH indicators
- Titration of an acid with a base
- Titrate an acid with a base of known molarity
- Calculate the pH of an acid and a base
- Inverting the pH equation to find concentration
- Find the concentration from the pH

- Buffers resist a change in pH
- Ionization constant for water
- Electrolytes and non-electrolytes
- Strong and weak electrolytes
- Total ionic equations
- Write a total ionic equation
- Net ionic equations
- Write a net ionic equation

13. What is Chemical Equilibrium?

- Forward and reverse reactions
- Chemical equilibrium
- Reaching chemical equilibrium
- Collision theory
- Energy of collisions affects reaction rate
- Frequency of collisions affects reaction rate
- Concentration of reactants affects reaction rate
- Distribution of reactants affects reaction rate
- Orientation of reactants affects reaction rate
- Catalysts and how they work
- Activation energy and transition point
- Exothermic and endothermic reactions
- General equilibrium constants
- Calculate a general equilibrium constant
- Heterogeneous equilibrium constants
- Calculate a heterogeneous equilibrium constant
- Ionization equilibrium constants
- Calculate an ionization equilibrium constant
- Solubility equilibrium constants
- Calculate a solubility equilibrium constant
- Le Chatelier's Principle
- Le Chatelier's Principle with concentration changes
- Le Chatelier's Principle with pressure changes
- Le Chatelier's Principle with temperature changes

14. What are Oxidation Numbers?

- Oxidation and reduction
- Redox reactions
- Oxidizing and reducing agents
- Determining oxidation numbers
- Calculate some oxidation numbers
- Half-reaction redox balancing method
- Balance an equation using half-reaction method
- Reduction potential
- Find the reduction potential
- Oxidation potential
- Find the oxidation potential
- Electrochemistry
- Electrolytic batteries
- Voltaic batteries

XIV

Chapter 1

Credit: http://www.pcgameshardware.com/?menu=browser&article_id=668994&image_id=679433

What is Matter?

A Message to the Student

This book makes you a promise: it will tell you only the Need-to-Know and skip all of the Nice-to-Know information. Although this will not waste your time with unnecessary stuff, it also makes it very, very important that you carefully read and understand every word. Why? Because ALL of it is Need-to-Know! So you may have to read each chapter (or at least parts of it) twice and sometimes even three times. Let's get started.

Matter and States of Matter

You probably already have an idea of what matter is. It is something you can hold in your hand, and it is at least a little bit heavy. What is the official definition of **matter**? It is anything that occupies space and has mass, which is the same idea but said in a more scientific (and confusing) way.

What is something that is not matter? If you think about it, you cannot hold light in your hand, and it does not weigh anything. So light is not matter.

Matter is the material from which all the universe is made. It comes in many forms, shapes, and combinations, and it often changes from one to the other. You need to recognize these forms, shapes, and combinations and how matter changes from one to the other. In particular, you need to recognize chemical changes because chemistry is the study of chemical changes.

Matter comes in three forms: **solid**, **liquid**, and **gas**. Actually there is a fourth one: **plasma**, sometimes also called fire. Chemistry does not deal with plasma, so it will be totally ignored in this book.

These are called the **physical forms** (also called **physical states**) of matter. They are called physical forms to tell them apart from the chemical forms of matter that you will learn about later in the chapter.

You can tell one physical form of matter from another by looking at its shape and volume. A 10 cubic centimeter (cc) ice cube (a solid) will remain a 10 cc ice cube regardless of where you put it. The ice cube will also keep its cubic shape as long as it remains a solid (frozen). So it has a fixed volume and a fixed shape. All solids are like that.

However, a 10 milliliter (ml) bottle of wine (a liquid), although it will remain 10 ml of wine no matter where you put it, will take the shape of the wine glass you put it in. So it has a fixed volume but not a fixed shape. It is said that its shape is **variable** (it varies). All liquids are like that.

Finally a 1000 cc balloon of helium (a gas) will become a 10,000 cc balloon of helium if you let it rise into the sky, and its shape will take the shape of a cylinder or whatever you put it in. So it does not have either a fixed volume or a fixed shape. Its shape and volume are variable. All gases are like that.

Also notice that a gas can be compressed a lot (from 10,000 cc to 1000 cc), while a liquid or a solid can be compressed only very little. Table 1-1 shows you the differences between the physical forms of matter.

What is Matter?

Table 1-1: The Forms of Matter

	Volume	Shape	Compressable
Solid	Fixed	Fixed	No
Liquid	Fixed	Variable	No
Gas	Variable	Variable	Yes

You can change matter from one physical form to another by heating or cooling it. If you heat an ice cube (a solid), it **melts** into water (a liquid), and if you keep on heating it, it **boils** into steam (a gas). If you then cool the steam (a gas), it **condenses** into water (a liquid), and eventually **freezes** into ice (a solid). There are a few exceptions, like carbon dioxide, which go directly from a solid to a gas (**sublimate**) without ever being a liquid. If you cool them, these gases go directly from a gas to a solid (**deposit**) without ever being a liquid. All these changes are called **physical changes** as opposed to the other chemical changes you will learn about later in the chapter.

You know that by cooling something, you are taking away heat. Heat is a form of energy. This means that solids have less energy than liquids, and liquids have less energy than gases. Without going into details (Nice-to-Know information), heat determines how fast the molecules vibrate. So it means solids vibrate less than liquids, and liquids vibrate less than gases.

Physical versus Chemical Changes

You have to wait until the end of the chapter to learn what a chemical change is. First you need to know what a chemical change is not.

Physical changes can sometimes be confused with chemical changes. For example, if you had never seen water freeze, you might think you were seeing a chemical change. After all, you see the water change from a clear liquid into a white solid. But you now know that it is a physical change.

If you take a block of ice and melt it, then boil it, you will still have the water you started with, only it will be steam instead of ice. But if you combine sodium with chlorine, on the other hand, you will get sodium chloride (ordinary table salt). Sodium and chlorine, as you probably know, are both very toxic (will kill you), but salt is not. The average human being eats at least 2 grams of salt every day and does not die from it. Can you see that the salt cannot be the same material (**substance**) that you started with?

This then is the major difference between physical and chemical changes. With a **physical change** you still have the same material you started with. Freezing water is an example. With a **chemical change** you have a different material from the ones you started with. Table salt is an example.

There is another important difference. Physical changes can be undone (reversed) by physical means (ways), like heating or cooling. Other chemical changes can only be reversed by chemical means (other chemical changes).

Mixtures and Compounds

Mixing two substances together can also be confused with a chemical change. Mixtures (air for example) can have a fixed ratio by weight (mass) of the substances they came from, just like compounds. So you need to take a close look at mixtures.

One way of combining substances is by mixing them. For example, salt can be mixed with pepper. Such a mixture can be separated by physical means. You could (if you really, really wanted to) take a magnifying glass and a pair of tweezers and separate the salt from the pepper one grain at a time. So mixing is a physical change because it can be reversed by physical means.

You usually think of a mixture as between two solids or two gases, but really mixtures come in all combinations of physical forms. For example, you can have a mixture between a liquid and a gas (carbonated soda), a liquid and a solid (dental fillings), and even a solid and a solid (a metal alloy).

There are two types of mixtures. **Heterogeneous** mixtures are not the same everywhere. For example, if you take a sample of dirt in southern Arizona, it will be mostly sand with a little granite. But if you take a sample of dirt in central Colorado, it will be mostly granite with a little sand. So topsoil is a heterogeneous mixture. A heterogeneous mixture is not confused with a chemical compound very often because a chemical compound is the same everywhere, and a heterogeneous mixture is not.

But **homogeneous** mixtures are the same everywhere. Air is a homogeneous mixture of nitrogen and oxygen. If you take a sample of air in Arizona and compare it to one in Colorado, they will both be 11% oxygen and 89% nitrogen by mass. Sea water is another homogenous mixture (salt and water). If you take a sample of sea water in the North Atlantic it will be exactly the same as one taken in the South Pacific.

But even homogenous mixtures can be separated by physical means. If you take a sample of air and cool it to -183 °C, the oxygen will liquefy (turn into a liquid), leaving only the nitrogen behind. Likewise, if you boil sea water, the water will evaporate, leaving only salt behind. So homogenous mixtures are physical changes because they can be reversed by physical means.

Elements and Compounds

Okay, so it isn't just melting or boiling something, and it isn't just mixing two things together. So what is it? What is a other chemical changes?

To answer that question you first need to know about the elements. An **element** is a pure (not combined with anything else) substance. It cannot be broken down into simpler materials, not even by chemical means. The very smallest part of an element is called an **atom**.

There are a little over 110 known elements, and new ones are being discovered about every ten years. The symbols for these elements are shown in the Periodic Table that you will find inside the front cover of this book. You will learn a lot more about the Periodic Table in Chapter 3.

Most of the chemical symbols for the elements start with the first letter of their name written as a capital letter. For example, C is the symbol for carbon, N is the symbol for nitrogen, and O is the symbol for oxygen. If you have two elements that start with the same letter, a lower case letter is added to tell them apart. For example, carbon is C, but chlorine is Cl, calcium is Ca, chromium is Cr, and cobalt is Co.

Ten of the elements were named during the Middle Ages when Latin was spoken in Europe. These elements start with the first letter of their Latin names. So iron is Fe, gold is Au, silver is Ag, mercury is Hg, sodium is Na, potassium is K, tin is Sn, antimony is Sb, tungsten is W, and lead is Pb.

There is a set of flash cards just inside the back cover of this book. You need to cut them out today and start memorizing the names and symbols of the most commonly used elements and their ions.

Compounds are made by combining two or more atoms. The very smallest part of a compound is called a **molecule**, not an atom, because it is made up of two or more atoms instead of only one. For example, sodium chloride (written NaCl) is a molecule made by chemically combining an atom of sodium (Na) with an atom of chlorine (Cl).

There are thousands of compounds because the 110+ elements can be combined in thousands of ways. Of course all of these compounds can be broken back down into the elements they came from by chemical means (ways) that you will learn about later on. For example you can break NaCl back into Na and Cl by chemical means.

The chemical **formula** for a compound is written as a combination of the symbols of the elements it is made of. You probably know that H_2O is the formula for water. What you may not know is that the subscript 2 means there are two hydrogen atoms in each molecule of water. The subscript 1 is never written. It is just assumed to be a 1 if there is no other number there. So water also has one oxygen atom in each molecule. Likewise the formula for sulfuric acid is H_2SO_4. That means there are two hydrogen atoms, one sulfur atom, and four oxygen atoms in every molecule of sulfuric acid.

Chemical Reactions

The forming of compounds or breaking them apart is what is called a **chemical change**, also called a **chemical reaction**. **Chemistry** is the study of chemical changes. You will learn a lot more about chemical reactions and how they happen in Chapter 4.

Branches of Chemistry

But before we get off that subject, you should know that there are two main branches of chemistry. **Organic chemistry** deals with compounds that contain **hydro-carbons** (have both H and C). **Inorganic chemistry** deals with compounds that don't contain hydro-carbons. Other branches of chemistry include biochemistry, nuclear chemistry, and physical chemistry. This book will deal (mostly) with inorganic chemistry.

Next you need to learn what an atom looks like.

Homework Problems

1. Which physical form has a definite volume and a variable shape?

    a. Gas

    b. Liquid

    c. Solid

    d. Plasma

    e. None of the above

2. Which physical form has a definite volume and a definite shape?

    a. Gas

    b. Liquid

    c. Solid

    d. Plasma

    e. None of the above

3. Which physical form has a variable shape and a variable volume?

    a. Gas

    b. Liquid

    c. Solid

    d. Plasma

    e. None of the above

4. Which physical form has a variable shape and cannot be compressed?

    a. Gas

    b. Liquid

    c. Solid

    d. Plasma

    e. None of the above

5. What do we call the following changes of physical form?

    a. Gas goes to a liquid

    b. Gas goes to a solid

    c. Solid goes to a gas

    d. Solid goes to a liquid

    e. Not applicable

What is Matter?   7

6. What do we call the following changes of physical form?

   a. Liquid goes to a gas

   b. Gas goes to a solid

   c. Gas goes to a liquid

   d. Liquid goes to a solid

   e. Not applicable

7. Are the following chemical or physical changes?

   a. Wood burning in a roaring fireplace

   b. Dry ice making fog during a rock concert

   c. Sugar dissolving in a cup of hot coffee

   d. An icemaker in your refrigerator making ice cubes

   e. Not applicable

8. Are the following chemical or physical changes?

   a. Rust forming on the fender of an old car

   b. A river freezing over in the winter

   c. Gasoline burning in your car's engine

   d. A cake baking in the oven

   e. Not applicable

9. What are the chemical symbols of the following elements?

   a. Nitrogen

   b. Tungsten

   c. Chlorine

   d. Silver

   e. Not applicable

10. What are the chemical symbols of the following elements?

    a. Oxygen

    b. Gold

    c. Helium

    d. Mercury

    e. Not applicable

11. What are the names of the following chemical symbols?

    a. Ne

    b. K

    c. Pt

    d. Pb

    e. Not applicable

12. What are the names of the following chemical symbols?

    a. Al

    b. W

    c. Ni

    d. Fe

    e. Not applicable

13. Sterling silver is an alloy of silver and gold. What is it?

    a. An element

    b. A compound

    c. A homogeneous mixture

    d. A heterogeneous mixture

    e. None of the above

14. Brass is an alloy of copper and zinc. What is it?

    a. An element

    b. A compound

    c. A homogeneous mixture

    d. A heterogeneous mixture

    e. None of the above

15. Carbon dioxide is made from carbon and oxygen. What is it?

    a. An element

    b. A compound

    c. A homogeneous mixture

    d. A heterogeneous mixture

    e. None of the above

16. Water is made from hydrogen and oxygen. What is it?

    a. An element

    b. A compound

    c. A homogeneous mixture

    d. A heterogeneous mixture

    e. None of the above

17. The edge of a beach is a mixture of sand and water. Which of these correctly describe it?

    a. A mixture of elements

    b. A mixture of compounds

    c. A homogeneous mixture

    d. A heterogeneous mixture

    e. None of the above

18. An artist's oil paint is a mixture of pigments and oil. Which of these correctly describe it?

    a. A mixture of elements

    b. A mixture of compounds

    c. A homogeneous mixture

    d. A heterogeneous mixture

    e. None of the above

19. Which of the following are elements?

    a. Cl

    b. MgO

    c. Co

    d. CO_2

    e. None of the above

20. Which of the following are compounds?

    a. Cl

    b. MgO

    c. Co

    d. CO_2

    e. None of the above

21. From which elements is water made?

    a. O and N

    b. S and H

    c. O and H

    d. S and N

    e. None of the above

22. From which elements is table salt made?

    a. Cl and Ni

    b. Co and Ni

    c. Co and Na

    d. Cl and Na

    e. None of the above

23. From which elements is carbon dioxide made?

    a. O and C

    b. N and D

    c. O and S

    d. D and C

    e. None of the above

24. From which elements is sulfur dioxide made?

    a. D and S

    b. O and S

    c. D and C

    d. O and C

    e. None of the above

Chapter 2

Credit: http://www.cosmosmagazine.com/files/imagecache/news/files/news/20090123_atom.jpg

What does an Atom Look Like?

In keeping with our promise not to waste your time, to skip all the Nice-to-Know and present only the Need-to-Know, we are going to leave out the history of how we learned all this. Instead we are just going to tell you what we know about what an atom looks like (its **structure**).

Protons, Electrons, and Neutrons

Atoms are made up of three types of particles. They are called: **protons,** which have a very small mass and a positive (+) charge; **electrons,** which have an even smaller mass and a negative charge (-); and **neutrons,** which have a mass almost the same as a proton but with no charge (0). The differences between these particles are shown in Table 2-1.

Table 2-1: Particles found in an Atom

	Mass (amu)	Size (adu)	Charge (1e)
Proton	1.0000	1.0000	+1
Electron	0.0005	0.0793	-1
Neutron	1.0014	1.0006	0

Charge is something about matter that is not really understood. But for learning chemistry it is enough that you know that a particle with a negative charge is **attracted** to (pulled towards) a particle with a positive charge. On the other hand, a particle with a negative charge and another particle with a negative charge **repel** (push away from) each other. The same is true for two particles with positive charges; they repel each other. So opposite charges attract but like ones repel each other.

The Nucleus

One or more protons and possibly one or more neutrons are in the **nucleus** (center) of an atom, and possibly one or more electrons orbital around the nucleus like the planets orbit around the sun. This overly simple picture of the atom is shown in Figure 2-2.

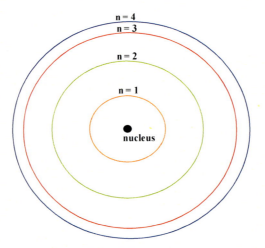

Figure 2-2: Orbits of an Atom

But unlike planets, the electrons cannot orbit any distance from the nucleus they want. Only orbits with certain distances from the nucleus are allowed. The reason for this falls into the Nice-to-Know category, which we will skip. It is enough that you know that only orbits at certain distances from the nucleus are possible.

Orbits and Orbitals

The first allowed orbit is given the number 1, the second is given the number 2, the third 3, the fourth 4, and so on. To make it a little more complicated, each main orbit has one or more sub-orbits called **orbitals**. These are given the names **s, p, d,** and **f**. So, for example, orbit number 4 has 4s, 4p, 4d, and 4f orbitals.

The orbit number tells you how many sub-orbits there are in that orbit. So orbit 1 has only one (1s), but orbit 2 has two (2s and 2p), orbit 3 has three (3s, 3p, and 3d), and so on. This slightly more complicated picture of the atom is shown in Figure 2-3.

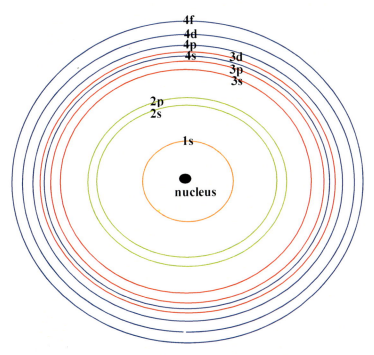

Figure 2-3: Orbitals of an Atom

Energy Levels of Orbitals

Notice that as the orbit number increases, the orbits get closer and closer together, and by the time you get to the fourth main orbit, their orbitals begin to overlap. So the 4s orbital is actually closer to the nucleus than the 3d, like shown in Figure 2-3. The same thing happens with the 5s orbital. It is closer to the nucleus than the 4d orbital. So the orbitals, in increasing distance from the nucleus, are **1s 2s 2p 3s 3p 4s 3d 4p 5s 4d 4f ...**, like shown in the drawing.

It takes work (energy) to move an electron from a lower orbital to a higher orbital because the positive protons in the nucleus and the negative electron attract each other, and you have to pull them apart. This means that the 2s orbital has more energy than the 1s orbital, so the 1s orbital is preferred by 2s electrons looking to have less energy. Likewise the 3d orbital has more energy than the 4s orbital, so the 4s orbital is preferred by 3d electrons looking to have less energy. What this means is that the order of increasing distance from the nucleus is the same as the order of increasing energy.

Excitation of Electrons and Emission of Photons

If a voltage is applied across a florescent tube (if you turn on the classroom lights), the voltage causes the electrons of the atoms in the florescent tube to move to a higher energy orbital. So the electrons are **excited** (kicked) into a higher orbital by the energy from the voltage across the florescent tube.

Electrons like to have the lowest possible energy. So after a fraction of a second, the electrons fall back down to the lower energy orbitals they came from. When that happens, they give off the extra energy by shooting off a unit of light called a **photon**. So the electron **emits** a photon. The energy of the photon is the difference in energy of the orbital the electron came from and the orbital it fell into. These photons are the light that comes from the florescent tubes.

Atomic Spectra

Photons vibrate, and photons of different energies vibrate at different frequencies. The color of the light you see depends on the **frequency** (number of times it vibrates per second) of that photon. The energy of a photon also depends on its frequency. So photons that are blue vibrate faster than red ones, and blue ones also have more energy than red ones.

Remember that the energy of a photon depends on the difference in energy of the orbital the electron came from and the orbital it fell into. Also remember that only certain orbitals are possible for electrons orbiting around a nucleus. That means that only certain energies are possible.

Since the energy of a photon depends on its frequency, and the color of the light emitted depends on its frequency, the fact that only certain energies are possible means only certain colors are seen in the light emitted by an atom.

The **spectrum** of an element is the collection of colors an atom emits from all the different orbitals possible for that atom. Different elements have different spectrums because different elements have different possible orbitals. So the spectrum of an element is sort of like a fingerprint of that element. The spectrums of several different elements are shown in Figure 2-4.

Electron Configurations

In orbitals, electrons want to be in pairs. In fact, one of them will always orbit clockwise while the other will orbit counter-clockwise. The reason for this is in the Nice-to-Know category, so don't worry about it. So the 1s orbital can have two electrons in it, one orbiting clockwise and the other counter-clockwise. Likewise all the other orbitals can have two electrons in them, one orbiting clockwise and the other counter-clockwise.

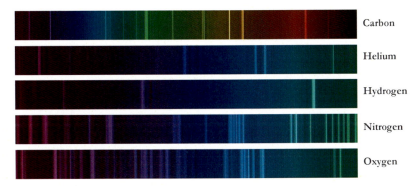

Figure 2-4: The Spectrums of Several Different Elements

Not all main orbits have only one orbital. In fact each level can have up to **one s** orbital, **three p** orbitals, **five d** orbitals, and **seven f** orbitals, depending on the orbit number. This is shown in Table 2-5. Why there is this one-three-five-seven sequence is not really understood and is Nice-to-Know anyway, so don't worry about it. It is enough that you know that an orbit can have one s, three p, five d, and seven f orbitals and that each of these orbitals can have two electrons in it. In the 4th orbit, for example, the 4s orbital can have 2 electrons, the three 4p orbitals can have 6 electrons, the five 4d orbitals can have 10 electrons, and the seven 4f orbitals can have 14 electrons, for a total of 32 electrons possible in orbit number 4 alone.

Table 2-5: Orbits and Orbitals of an Atom

Main Orbit Number	Orbitals Present	Maximum Electrons
1	1s	2
2	2s	2
	2p	6
3	3s	2
	3p	6
	3d	10
4	4s	2
	4p	6
	4d	10
	4f	14

The **electron configuration** of an atom with 35 electrons (bromine) is written $1s^2\ 2s^2\ 2p^6\ 3s^2\ 3p^6\ 4s^2\ 3d^{10}\ 4p^5$. Notice that the order of the orbitals is the same as the order of increasing energy like before and that the number of electrons in each orbital is shown as a superscript. Also notice that s orbitals have 2 electrons, p orbitals have 6 electrons, d orbitals have 10 electrons, and that the sum of all of the electrons is 35. The last orbital (4p) has only 5 electrons in it (not 6) because we ran out of electrons.

So how do you know how many electrons an atom has? Well the number of electrons is the same as the number of protons in that atom, and the number of protons for each element is shown in the Periodic Table above the symbol for that element. Look at the Periodic Table (inside the front cover). Can you see that bromine has the symbol Br and has a 35 above its symbol? So bromine has 35 protons, and that means it has 35 electrons. You will learn more about the Periodic Table in Chapter 3.

Positive and Negative Ions

Ordinarily an atom has no net charge. That means that the number of protons (plus charges) equals the number of electrons (minus charges). If you add up the plus and minus charges, you get a net charge of zero.

But that isn't always the case. An atom sometimes loses one or more electrons. When that happens it becomes a **positive ion**. It then has more protons than it has electrons. Since protons are positive and electrons are negative, the ion is a little more positive than it is negative.

Likewise, an atom sometimes gains one or more electrons. When that happens it becomes a **negative ion**. It then has fewer protons than it has electrons. Since protons are positive and electrons are negative, the ion is a little more negative than it is positive.

Positive ions are written with a + superscript over their symbol, like Na^+. If an atom loses two electrons it is written with a +2 superscript, like Mg^{+2}; if it loses three electrons, with a +3 superscript, like Al^{+3}; and so on.

Negative ions are written with a − superscript over their symbol, like Cl^-. If an atom gains two electrons it is written with a -2 superscript, like S^{-2}; if it gains three electrons, with a -3 superscript, like P^{-3}; and so on.

When you write the electron configurations for positive or negative ions, you need to add or subtract electrons to take into account the extra electrons. For example, the configuration for Ca (work it out) is $1s^2\ 2s^2\ 2p^6\ 3s^2\ 3p^6\ 4s^2$. For Ca^{+2}, it is $1s^2\ 2s^2\ 2p^6\ 3s^2\ 3p^6$, with two less electrons. Likewise for O it is $1s^2\ 2s^2\ 2p^4$. For O^{-2}, it is $1s^2\ 2s^2\ 2p^6$, with two more electrons.

The configuration for Ne (work it out) is $1s^2\ 2s^2\ 2p^6$, exactly the same as that for O^{-2}. In this case the O^{-2} ion is said to be **isoelectronic** with Ne. That's just a fancy word that means that their electron configurations look alike.

That idea gives us an abbreviation for writing electron configurations. If you work it out, the configuration for Mg is $1s^2 2s^2 2p^6 3s^2$ and the configuration for Ne is $1s^2 2s^2 2p^6$. So you can abbreviate the configuration for Mg as [Ne] $3s^2$, where [Ne] is the configuration for Ne. Abbreviations like that can save you lots of time writing configurations for atoms with many electrons.

Atomic Notation

One more thing you need to know about atoms is **atomic notation.** It tells you how many protons and neutrons are in the nucleus of an element and is written like this

What does an Atom Look Like?

$$^{m}_{n}X$$

where X is the symbol of the element, n is the **atomic number** (number of protons) and m is the **mass number** (number of protons plus neutrons). If **m** is the number of protons plus neutrons and **n** is the number of protons, can you see that **m** minus **n** must be the number of neutrons in the nucleus?

As an example, the atomic notation for carbon twelve (written C-12) is

$$^{12}_{6}C$$

because it has 6 protons (look up carbon in the Periodic Table) and 12 protons plus neutrons (that's why it's called C-12). Since 12 minus 6 is equal to 6, C-12 has 6 neutrons in its nucleus.

You may know that there is also a carbon fourteen (C-14) that is used by archeologists to date very old artifacts. The atomic notation for C-14 is

$$^{14}_{6}C$$

because it has 6 protons (otherwise it wouldn't be carbon) and 14 protons plus neutrons. Since 14 minus 6 is equal to 8, C-14 has 8 neutrons in its nucleus. Atoms like C-12 and C-14 that have the same number of protons but different number of neutrons are called **isotopes**.

Because isotopes exist in nature with different percentages and have different atomic weights, if you need to know the weight of an average sample of carbon, you have to multiply the **natural abundance** (percentage found in nature) of C-12 times its weight (12) and add to that number the natural abundance of C-14 times its weight (14).

Carbon has 98.8% C-12, so 0.988 times 12 is 11.86 **amu** (atomic mass units). Carbon also has 1.1% C-14, so 0.011 times 14 is 0.15 amu. When you add them together, you get 12.01 amu. If you look at the Periodic Table, that is exactly the number shown below the **C**. It is called the **atomic mass** for that element and is the weight of an average sample of carbon.

Two Myths

Before we get off the subject of atoms, there are two common myths about them that come from an old theory on atoms called the Dalton theory.

The first myth says all atoms of an element are identical (the same). But as you just saw, there are two isotopes of carbon, C-12 and C-14. They are both carbon but have different numbers of neutrons. So they are not identical.

The second myth says all atoms are indivisible (cannot be broken up). But as you probably know, nuclear power stations get huge amounts of energy by splitting uranium atoms into smaller ones. So atoms are not indivisible.

Now let's take a close look at the Periodic Table.

Homework Problems

1. Which the following are not found in the nucleus of an atom?

    a. Proton

    b. Electron

    c. Photon

    d. Neutron

    e. None of the above

2. Which of the following are found in the nucleus of an atom?

    a. Proton

    b. Electron

    c. Photon

    d. Neutron

    e. None of the above

3. What is the minimum number of protons found in a nucleus?

    a. Two

    b. Three

    c. One

    d. None

    e. None of the above

4. What is the minimum number of neutrons found in a nucleus?

    a. Two

    b. Three

    c. One

    d. None

    e. None of the above

5. How many sub-orbits (orbitals) does the main orbit 3 have?

    a. Two

    b. Three

    c. One

    d. None

    e. None of the above

What does an Atom Look Like?

6. How many sub-orbits (orbitals) does the main orbit 2 have?

   a. Two

   b. Three

   c. One

   d. None

   e. None of the above

7. How many electrons does zinc have?

   a. 15

   b. 22

   c. 30

   d. 37

   e. None of the above

8. How many electrons does phosphorus have?

   a. 15

   b. 22

   c. 30

   d. 37

   e. None of the above

9. How many protons, neutrons, and electrons does carbon-14 have?

   a. 6,8,6

   b. 6,6,8

   c. 8,6,6

   d. 6,6,6

   e. None of the above

10. How many protons, neutrons, and electrons does beryllium-9 have?

    a. 4,5,4

    b. 4,4,4

    c. 5,4,5

    d. 4,5,6

    e. None of the above

11. What element has the electron configuration $1s^2, 2s^2, 2p^6, 3s^2, 2p^1$?

    a. Na

    b. Al

    c. Ni

    d. Cu

    e. None of the above

12. What element has this structure $1s^2, 2s^2, 2p^6, 3s^2, 2p^6, 3s^2, 3p^6, 4s^2, 3d^1$?

    a. Na

    b. Al

    c. Ni

    d. Cu

    e. None of the above

13. Which main orbit has a d orbital but not an f orbital?

    a. n=4

    b. n=1

    c. n=3

    d. n=2

    e. None of the above

14. Which main orbit has a d orbital and an f orbital?

    a. n=4

    b. n=1

    c. n=3

    d. n=2

    e. None of the above

15. Which main orbit has only an s orbital?

    a. n=4

    b. n=1

    c. n=3

    d. n=2

    e. None of the above

What does an Atom Look Like?

16. Which main orbit has a p orbital but neither a d orbital or an f orbital?

   a. n=4

   b. n=1

   c. n=3

   d. n=2

   e. None of the above

17. Which of the following is the electron configuration for cobalt?

   a. $1s^2, 2s^2, 2p^6, 3s^2, 2p^6, 3s^2, 3p^6, 4s^2$

   b. $1s^2, 2s^2, 2p^6, 3s^2, 2p^6, 3s^2, 3p^6, 4s^2, 3d^1$

   c. $1s^2, 2s^2, 2p^6, 3s^2, 2p^6, 3s^2, 3p^6$

   d. $1s^2, 2s^2, 2p^6, 3s^2, 2p^6, 3s^2, 3p^6, 4s^1$

   e. None of the above

18. Which of the following is the electron configuration for potassium?

   a. $1s^2, 2s^2, 2p^6, 3s^2, 2p^6, 3s^1$

   b. $1s^2, 2s^2, 2p^6, 3s^2, 2p^5$

   c. $1s^2, 2s^2, 2p^6, 3s^2, 2p^6, 3s^2$

   d. $1s^2, 2s^2, 2p^6, 3s^2, 2p^1$

   e. None of the above

19. If there were only two isotopes of lithium, L-6 (10%) and L-7 (90%), what would be the mass shown under Li in the Periodic Table?

   a. 6.3

   b. 6.9

   c. 7.2

   d. 7.6

   e. None of the above

20. If there were only two isotopes of hydrogen, H-1 (90%) and H-2 (10%), what would be the mass shown under H in the Periodic Table?

   a. 1.3

   b. 0.7

   c. 1.1

   d. 0.9

   e. None of the above

21. If there were only two isotopes of uranium, U-235 (10%) and U-238 (90%), what would be the mass shown under U in the Periodic Table?

    a. 235.3

    b. 233.9

    c. 239.6

    d. 237.7

    e. None of the above

22. If there were only two isotopes of lead, Pb-207 (10%) and Pb-208 (90%), what would be the mass shown under Pb in the Periodic Table?

    a. 208.3

    b. 207.9

    c. 208.6

    d. 207.5

    e. None of the above

23. Draw the helium atom showing electrons, protons, and neutrons.

24. Draw the beryllium atom showing electrons, protons, and neutrons.

Chapter 3

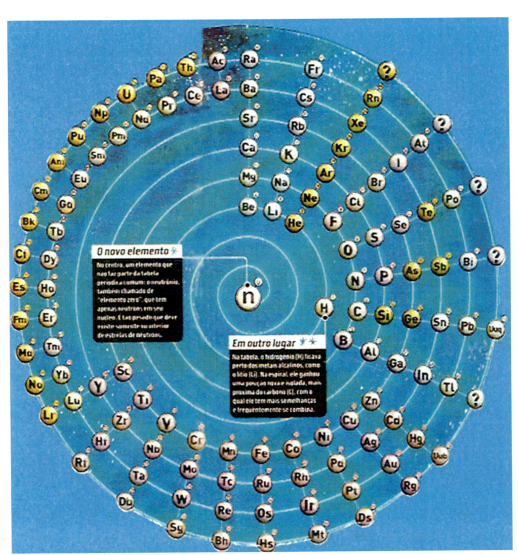

Credit: http://4.bp.blogspot.com/_SKKSfIMaDAc/TCYsmAYXbaI/AAAAAAAABg/N7GIl2mtM-8/s1600/Nova+Tabela+Peri%C3%B3dica.gif

What is the Periodic Table?

The short answer to that question is that it's the table inside the front cover of this book. But that doesn't tell you how to read it or what it can tell you. You need to know that. That's exactly what you're going to learn now!

Information about Elements

Each of the boxes in the Periodic Table is one of the elements. You already know that the letters in the middle are abbreviations for the names of the elements. Just above the name of each element is the **atomic number**. It tells you how many protons that element has. If you look closely you will see that the atomic number increases by one as you go from left to right in the Periodic Table. Just below the name of each element is the **atomic mass**. It tells you how much an average sample of that element weighs. Do you remember that it is the weighted average of the isotopes of that element?

At room temperature, most of the elements are solids, but two of them (Hg and Br) are liquids, and eleven of them (H_2, N_2, O_2, F_2, Cl_2, He, Ne, Ar, Kr, Xe, and Rn) are gases. Notice that the first five gases are in pairs. These gases are called **diatomic** because they are usually seen as molecules with two atoms of the same element bonded together. In this Periodic Table the symbols for solids are shown in black, the symbols for liquids are shown in blue, and the symbols for gases are shown in white.

Organization and Collective Names

The Periodic Table is arranged in 18 columns called **groups** and 7 rows called **series**. A group is sometimes also called a **family**, and a series is sometimes also called a **period**. The groups are numbered 1 to 18 across the top of the Periodic Table, and the series are numbered 1 to 7 down the left side of the Periodic Table. Groups are also sometimes numbered with Roman numerals like VII and the letters A and B. This is an old (and very confusing) way of numbering groups. This book will ignore the old system and use only the new system (group numbers 1 to 18) used by **IUPAC** (International Union of Pure and Applied Chemistry).

Some of the groups have **collective names**. You need to know them because chemists often talk about collections of elements by those names. The group on the far left side (group 1) is called the alkali metals. Just to the right of the alkali metals is the **alkaline earth metals**. On the far right side (group 18) is the **noble gases**. Just to the left of the noble gases is the **halogens**.

Likewise, two of the series also have collective names. These are at the bottom of the page below the rest of the Periodic Table. The top series is called the **lanthanides** because they follow lanthanum (La) in the Periodic Table. The bottom series is called the **actinides** because they follow actinium (Ac) in the Periodic Table. They are placed at the bottom instead of where they belong so as to make the Periodic Table more compact.

Finally there are names for blocks of groups. Those groups in the "valley" in the middle of the Periodic Table are called the **transition elements**. Those groups to the left and right of the "valley" are called the **representative elements**. And those groups at the bottom are called the **inner transition elements**. Figure 3-1 shows all those confusing names.

What is the Periodic Table?

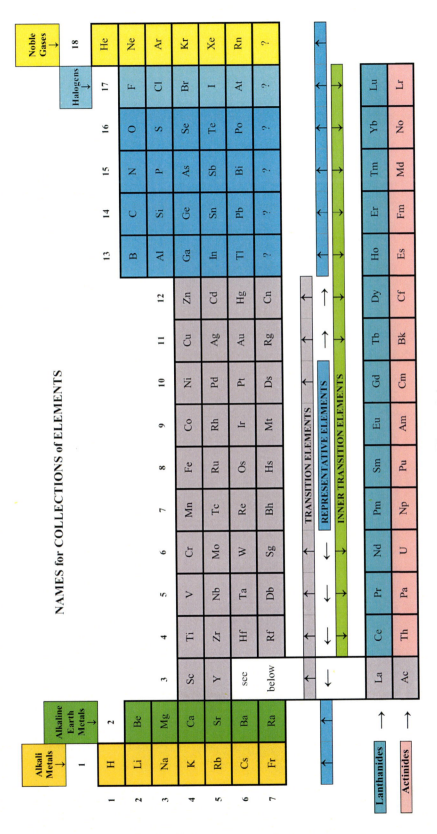

Figure 3-1

Information about Groups

The elements are lined up in groups by the way they make compounds with oxygen. Group 1 makes compounds with oxygen like X_2O where X stands for any element in that group and O is the symbol for oxygen. For example, water is H_2O, lithium oxide is Li_2O, sodium oxide is Na_2O, and so on. Group 2 makes compounds with oxygen like XO. So beryllium oxide is BeO, magnesium oxide is MgO, calcium oxide is CaO, and so on. Can you see the patterns?

The reason why you see these patterns has to do with the similar electron configurations of the elements in each group. Remember those $1s^2 2s^2 2p^6$... electron configurations? You will learn more about that later in this chapter.

Trends across the Periodic Table

The elements on the left side of the Periodic Table are called **metals**. Most of the elements (75%) are metals. The elements toward the upper right corner are called **non-metals**. They try to steal electrons from the metals. Between them, the elements angling down from boron (B) to astatine (At), along with some elements above and below that line, are called **semi-metals** or **semi-conductors**. The difference in chemical behavior between metals, non-metals, and semi-metals is called their **metallic character**.

The first element, hydrogen (H), is a little odd. Sometimes it acts like a metal, and sometimes it acts like a non-metal. In the Periodic Table it is put on the left side with the metals because it acts like a metal most of the time.

One more thing you should know about metallic character is that when you write a chemical formula, the metal always comes first and the non-metal second. So the compound made from sodium and chlorine is not ClNa, but NaCl, with the metal first and the non-metal second.

Many other properties of the elements follow the same pattern as metallic character. For example, the elements with the largest atomic diameter are in the lower left corner, and those with the smallest diameter are in the upper right corner. That's because metals hold onto their electrons only loosely, while non-metals hold onto their electrons very tightly. That makes the diameter of metal atoms larger than the diameter of non-metal atoms.

There is another important property of the elements that follows the same pattern. It is called **electro-negativity** (EN). That's just a polite word for greed, except that in the world of atoms, it's greed for electrons, not money. If you leave out the noble gases (you will see in Chapter 4 why the noble gases are left out when it comes to EN), the greediest elements (highest EN) are in the upper right corner of the Periodic Table, and those that are least greedy (lowest EN) are in the lower left corner. The less greedy metals often lose their electrons to the more greedy non-metals. That's why it's said that the non-metals steal electrons from the metals.

What is the Periodic Table?

METALLIC CHARACTER of ELEMENTS

Legend:
- Metals (blue)
- Semimetals (yellow)
- Nonmetals (orange)

	1	2	3	4	5	6	7	8	9	10	11	12	13	14	15	16	17	18
1	H																	He
2	Li	Be											B	C	N	O	F	Ne
3	Na	Mg											Al	Si	P	S	Cl	Ar
4	K	Ca	Sc	Ti	V	Cr	Mn	Fe	Co	Ni	Cu	Zn	Ga	Ge	As	Se	Br	Kr
5	Rb	Sr	Y	Zr	Nb	Mo	Tc	Ru	Rh	Pd	Ag	Cd	In	Sn	Sb	Te	I	Xe
6	Cs	Ba	La	Hf	Ta	W	Re	Os	Ir	Pt	Au	Hg	Tl	Pb	Bi	Po	At	Rn
7	Fr	Ra	Ac	Rf	Db	Sg	Bh	Hs	Mt	Ds	Rg	Cn	?	?	?	?	?	?

	Ce	Pr	Nd	Pm	Sm	Eu	Gd	Tb	Dy	Ho	Er	Tm	Yb	Lu
	Th	Pa	U	Np	Pu	Am	Cm	Bk	Cf	Es	Fm	Md	No	Lr

Figure 3-2

Trends down the Groups

Within each group, properties of the elements like atomic mass (weight), **density** (the mass of an element divided by its volume), melting point, and the like increase or decrease pretty evenly from one element to the next. So if you know the atomic mass of two elements in a group, you can guess at the atomic mass of another element in the same group.

For example, the atomic mass of Li is 6.94 and the atomic mass of Na is 22.99. Because the difference between them is 16.05, you would guess that K, the next element in group 1, has an atomic mass that is 16.05 higher than 22.99, or 39.04. The real value is 39.10. Not all predictions turn out that good, but they are usually close. You can make the same sort of predictions for other things like density, melting point, boiling point, and so on.

Electron Configurations and the Periodic Table

Remember from Chapter 2 that s orbitals can have (up to) 2 electrons, p orbitals 6 electrons, d orbitals 10 electrons, and f orbitals 14 electrons? Now if you place He on the left side next to H (He is usually put in group 18 because it acts like a noble gas, but it really belongs in group 2), you make a block 2 groups wide. If you look at the rest of the representative elements on the right side of the Periodic Table, they make a block 6 groups wide. The transition elements in the "valley" in the middle of the Periodic Table make a block 10 groups wide, and the inner transition elements at the bottom make a block 14 groups wide. Do you think that's just a coincidence?

So the Periodic Table is sort of a shadow of the electron configurations of the atoms. Why is that? Well in series 2 the configuration for B (figure it out) is 1s2, 2s2, 2p1; for C it is $1s^2$, $2s^2$, $2p^2$; for N it is $1s^2$, $2s^2$, $2p^3$; for O it is $1s^2$, $2s^2$, $2p^4$; for F it is $1s^2$, $2s^2$, $2p^5$; and for Ne it is $1s^2$, $2s^2$, $2p^6$. Can you see each has one more electron in the p orbital than the one just before?

So the answer is that groups 13 to 18 are filling p orbitals, groups 1 and 2 are filling s orbitals, groups 3 to 12 are filling d orbitals, and the two series at the bottom (they don't have group numbers) are filling f orbitals. You will see in Chapter 4 that this also explains why elements in the same group make compounds with oxygen that are similar.

The Periodic Law

If you put the elements in order of increasing number of protons in their nucleus, the chemical behavior of the elements falls into a repeating pattern. That statement is known as the **Periodic Law**. It is the reason way the Periodic Table is set up the way it is.

What is the Periodic Table?

Homework Problems

1. What is the symbol for the element in group 4 and series 6?
   a. Cr
   b. Cs
   c. Hf
   d. Ti
   e. None of the above

2. What is the symbol for the element in group 5 and series 7?
   a. Fr
   b. Db
   c. V
   d. Tc
   e. None of the above

3. What is the symbol for the element in group 7 and series 5?
   a. Nb
   b. Db
   c. Tc
   d. Bh
   e. None of the above

4. What is the symbol for the element in group 18 and series 3?
   a. Ne
   b. Li
   c. Cl
   d. He
   e. None of the above

5. What family is Co in?
   a. 9
   b. 6
   c. 4
   d. 11
   e. None of the above

6. What period is Au in?

   a. 5

   b. 6

   c. 8

   d. 13

   e. None of the above

7. Which of the following are metals?

   a. H

   b. O

   c. Ca

   d. I

   e. None of the above

8. Which of the following are non-metals?

   a. Ti

   b. Cl

   c. Ba

   d. N

   e. None of the above

9. Which of the following are semi-metals?

   a. Si

   b. Pb

   c. Ge

   d. Ga

   e. None of the above

10. Which of the following are semi-conductors?

    a. Se

    b. Pb

    c. Cl

    d. Sb

    e. None of the above

What is the Periodic Table?

11. Which of the following are transition elements?

   a. Mo

   b. Ca

   c. Hg

   d. Ce

   e. None of the above

12. Which of the following are inner transition elements?

   a. Ni

   b. Ce

   c. Eu

   d. Mo

   e. None of the above

13. Which of the following are representative elements?

   a. Fe

   b. Te

   c. Pt

   d. Ba

   e. None of the above

14. Which of the following are representative elements?

   a. Cr

   b. Si

   c. Hg

   d. Sr

   e. None of the above

15. Which of the following are actinides?

   a. U

   b. Cm

   c. Eu

   d. Ce

   e. None of the above

16. Which of the following are lanthanides?

    a. U

    b. Cm

    c. Eu

    d. Ce

    e. None of the above

17. Which of the following are halogens?

    a. I

    b. Xe

    c. Cl

    d. Ar

    e. None of the above

18. Which of the following are alkali earth metals?

    a. Na

    b. Be

    c. Cs

    d. Sr

    e. None of the above

19. Which of the following are alkali metals?

    a. Li

    b. Cs

    c. Ra

    d. Mg

    e. None of the above

20. Which of the following are noble gases?

    a. Cl

    b. F

    c. Ne

    d. Kr

    e. None of the above

21. What is the atomic number of Zr?
    a. 30
    b. 13
    c. 28
    d. 40
    e. None of the above

22. What is the atomic number of Pb?
    a. 78
    b. 82
    c. 46
    d. 84
    e. None of the above

23. What is the atomic mass of N?
    a. 12.01
    b. 9.01
    c. 14.01
    d. 1.01
    e. None of the above

24. What is the atomic mass of He?
    a. 19.00
    b. 52.00
    c. 16.00
    d. 4.00
    e. None of the above

Chapter 4

Credit: *http://blogs.wvgazette.com/karinfuller/files/2010/12/tug-of-war.jpg*

What is a Chemical Bond?

Chemical change, chemical combination, chemical reaction, chemical bond, these all mean more or less the same thing. Let's find out what that is.

Chemical Reactions

What is a **chemical bond?** The official definition is two atoms sharing one or more electrons. In simpler words, it's two atoms fighting over an electron.

It's like an old fashion tug-of-war. The team on the left pulls the rope to the left and the team on the right pulls the rope to the right. In doing so, the two teams pull on each other. Here the two teams are two atoms, and the rope they pull on is an electron. Can you see how this tug-of-war between the two atoms holds them together?

Under certain conditions chemical bonds can be formed (made) or broken apart. When that happens, we say that a **chemical reaction** has happened.

The new substance formed by a chemical reaction is called a **compound**. It is written with the symbols for the atoms right next to each other. Written like that it is called a chemical formula. For example, NaCl is the **chemical formula** for the compound made by a chemical reaction between Na and Cl.

Covalent Bonds

When an electron is shared equally (50-50) by two atoms, the bond is called a **covalent bond**. That means that one of the atoms has the electron 50% of the time and the other atom has it the other 50% of the time. Even a 40-60 (40% - 60%) sharing of electrons is considered (thought of as) a covalent bond. The actual dividing line between a covalent bond and other types of bonds is found by comparing the electro-negativity (EN) of the atoms involved. If it is less than or equal to 0.5, it is considered a covalent bond. The EN for most of the elements is shown in Figure 4-1.

Some good examples of covalent bonds are the **diatomic molecules** (H_2, N_2, O_2, F_2, Cl_2, Br_2, I_2 and Hg_2). They exist in pairs that are bonded together. Here the EN is the same for the two atoms because they are the same atom. So the difference in EN is zero, and they share their electrons 50-50.

As said before, even bonds between atoms that have different, but nearly equal, EN are considered covalent. Take nitrogen monoxide (NO) as an example. The EN for oxygen (O) is 3.5, and the EN for nitrogen (N) is 3.0. The difference in their EN is $3.5 - 3.0 = 0.5$, but the bond is still considered a covalent bond. Here the sharing of electrons is more like 40-60.

In the case of a compound held together by a covalent bond, the smallest piece that still is that compound is called a **molecule**.

What is a Chemical Reaction?

Electronegativities of Stable Elements

1	2											13	14	15	16	17	18
1 H 2.1																	2 He
3 Li 1.0	4 Be 1.5											5 B 2.0	6 C 2.5	7 N 3.0	8 O 3.5	9 F 4.0	10 Ne
11 Na 0.9	12 Mg 1.2	3	4	5	6	7	8	9	10	11	12	13 Al 1.5	14 Si 1.8	15 P 2.1	16 S 2.5	17 Cl 3.0	18 Ar
19 K 0.8	20 Ca 1.0	21 Sc 1.3	22 Ti 1.5	23 V 1.6	24 Cr 1.6	25 Mn 1.5	26 Fe 1.9	27 Co 1.8	28 Ni 1.8	29 Cu 1.9	30 Zn 1.6	31 Ga 1.6	32 Ge 1.8	33 As 2.0	34 Se 2.4	35 Br 2.8	36 Kr
37 Rb 0.8	38 Sr 1.0	39 Y 1.2	40 Zr 1.4	41 Nb 1.6	42 Mo 1.8	43 Tc 1.9	44 Ru 2.2	45 Rh 2.2	46 Pd 2.2	47 Ag 1.9	48 Cd 1.7	49 In 1.7	50 Sn 1.8	51 Sb 1.9	52 Te 2.1	53 I 2.5	54 Xe
55 Cs 0.7	56 Ba 0.9	57 La 1.1	72 Hf 1.3	73 Ta 1.5	74 W 1.7	75 Re 1.9	76 Os 2.2	77 Ir 2.2	78 Pt 2.2	79 Au 2.4	80 Hg 1.9	81 Tl 1.8	82 Pb 1.9	83 Bi 1.9	84 Po 2.0	85 At 2.2	86 Rn
87 Fr 0.7	88 Ra 0.9	89 Ac 1.1	104 Rf	105 Db	106 Sg	107 Bh	108 Hs	109 Mt	110 Ds	111 Rg	112 Cn	113 ?	114 ?	115 ?	116 ?	117 ?	118 ?

LEGEND

→ Atomic Number
→ Symbol
→ ELECTRONEGATIVITY

Figure 4-1

Polar Covalent Bonds

When the difference in EN is more than 0.5 (less than 40-60 sharing), you have instead a **polar covalent bond**. The molecule formed in this case is said to be **polarized** (have electric poles similar to the North and South poles of a magnet) because the electron spends a lot more of its time with one atom than it does with the other. So one atom is (on average) more negative than the other one. This is shown by what is called **delta notation**, using the Greek letter delta (δ) to mean a fraction of a positive or negative charge. For example, the bond in carbon monoxide (CO) is shown like this

$$^{\delta+}C - O^{\delta-}$$

In the case of a polar covalent compound, the smallest piece of the compound that still is that compound is called a **polarized molecule**.

Ionic Bonds

At the other end of the electron sharing range, when the difference in EN gets to be more than about 1.5, you have an **ionic bond**. Here an electron is almost completely passed from one atom to another (1-99 sharing). The sharing can be as high as 10-90 and still be considered an ionic bond.

In this case you show the charges by a full plus (+) and minus (-) because the atoms have almost completely lost or gained an electron. For example, the bond in sodium chloride is ionic. So the NaCl bond is shown like this

$$Na^+ - Cl^-$$

Because the electron was almost completely transferred from one atom to the other, the compound is now called a formula unit instead of a polarized molecule. So the smallest piece of an ionic compound that still is that compound is called a **formula unit.**

Valence Electrons

Not all electrons can be shared. Usually only the electrons in the outermost s and p orbitals can be shared. These electrons are called **valence electrons**.

How can you tell how many valence electrons an atom has? Well you could write out the electron configuration and count the number of electrons in the outermost s and p orbitals. But the IUPAC group number makes it easier. It so happens that if you look at the last digit of the group number, it tells you the number of valence electrons for an atom of any element in that group. For example an atom in group 16 has 6 valence electrons. Likewise an atom in group 3 has 3 valence electrons.

Notice that the atoms in group 18 (the noble gases) have 8 valence electrons, so their outermost s and p orbitals are completely filled. They have no place to put another electron and are said to have **closed shells**. Because of their closed shells, noble gases can't form compounds. They are said to be **inert** (non-reactive). By the way, that's why the noble gases have no electronegativity. They have all the electrons they can take.

What is a Chemical Reaction?

It turns out that all atoms want to be like noble gases and have closed shells. They will steal or give away electrons in order to become like them. This desire to have 8 valence electrons is called the **octet rule**. Atoms follow this rule when they form compounds. That means they arrange themselves in such a way that each one winds up with 8 valence electrons.

The only exception is hydrogen (H). Because it has no p orbital (only a 1s), H can have only 2 electrons. So it follows its own special **duet rule**.

Also notice that groups 9 and 10 are an exception to the last-digit-in-group-number rule. No atom can have less than 1 or more than 8 valence electrons. So the elements in groups 9 and 10 are considered to have 8 valence electrons even though the last digit of the group number says different.

Electron Dot Formulas

A neat way of showing valence electrons is using **electron dot formulas.** You first figure out the number of valence electrons in an atom. Then, starting at 3 o'clock and going clockwise, you put a dot for each valence electron next to its symbol, at 3, 6, 9, and 12 o'clock. So boron, which has 3 valence electrons, is shown this way

$$\cdot \overset{}{\underset{\cdot}{B}} \cdot$$

If you go all the way around and back to 3 o'clock, you put another dot there and so on. You keep going around like that until you run out of electrons. So sulfur, which has 6 valence electrons, is shown like this

$$\cdot \overset{\cdot}{\underset{\cdot\cdot}{S}} :$$

Lewis Dot Structures

The same idea can be used for compounds. When that is done, they are called **Lewis dot structures**. The procedure is a little different because there are more atoms involved. First find the number of valence electrons for each atom like you did before. Then add them all together and divide by two. The number you get is called the number of **electron pairs** (EPs).

Next you put the largest atom in the middle, and put four of the EPs around it. You then attach each of the other atoms to one of these four EPs. Finally, you divide up the remaining EPs among the atoms such that each one satisfies the octet (or duet) rule.

As an example, H_2S has two H atoms each of which has 1 valence electron. It also has an S atom which has 6 valence electrons. The total is 2 + 6 = 8. Divide by two and you get 4 EPs. The dot formula you get looks like this

$$H : \overset{\cdot\cdot}{\underset{\cdot\cdot}{S}} : H$$

Here the S atom has 8 electrons, and each of the H atoms shares 2 electrons with S. So the S satisfies the octet rule, and the two H satisfy the duet rule.

These dot structures can be thought of as a kind of picture of the chemical bonding that's going on between the atoms in a molecule. Here each of the H atoms is sharing its one electron with the S atom, and the S atom is sharing one of its 6 electrons with each of the H atoms. Can you see that?

Structural Formulas

Most chemists use a line in place of the pair of electrons that is bonding two atoms. When done that way, the picture you get is called a **structural formula**. Here it is for H_2S

$$H - \ddot{\underset{..}{S}} - H$$

Notice that the remaining non-bonding EPs are still shown as pairs of dots.

Sometimes you wind up with a dot formula that does not have enough EPs for every atom to satisfy the octet rule. When that happens, you need to move a pair of non-bonding electrons between the two atoms.

For example, following the steps above, you get this dot formula for CO_2

$$\ddot{\underset{..}{O}} : \ddot{\underset{..}{C}} : \ddot{\underset{..}{O}}$$

Here the C atom satisfies the octet rule because it has 8 electrons, but each of the O atoms has or shares only 6 electrons. In order to satisfy the octet rule, you need to move the C atom's top non-bonding EP between it and the O atom on the right and its bottom non-bonding EP between it and the O atom on the left. The dot formula now looks like this

$$\ddot{\underset{..}{O}} :: C :: \ddot{\underset{..}{O}}$$

Now each of the O atoms has or shares 8 electrons and satisfies the octet rule. The C atom also shares 8 electrons, so it satisfies the octet rule too. When you replace each of the bonding EPs with a line like before, you get this structural formula

$$\ddot{O} = C = \ddot{O}$$

The double lines are called **double bonds**. CO_2 has two double bonds.

You can even have a **triple bond**, like in N_2. In that case, you need to move three non-bonding EPs between the two N atoms to satisfy the octet rule. Its structural formula has three lines between the two atoms.

Polyatomic Ions

The dot formulas can also be used with polyatomic ions. They are called **polyatomic** because they have more than one atom. They are called **ions** because they have a plus or minus charge on them.

For example, ClO_2 has a total of $7 + 6 + 6 = 19$ valence electrons. If you divide that by 2, you get 9½ EPs. It can't have a half EP. In order to have a whole number of EPs, it needs to have an even number of electrons; and to have that, it has to either lose an electron or steal one. Cl being very electro-negative will steal one from an unsuspecting nearby atom. When you add in the extra electron, it now has 20 valence electrons (10 EPs) and is ClO_2^-. Its dot formula is

$$\left[:\overset{..}{\underset{..}{O}}:\overset{..}{\underset{..}{Cl}}:\overset{..}{\underset{..}{O}}: \right]^-$$

Notice that the dot formula for any polyatomic ion has to have brackets around it with its charge outside the brackets.

Likewise, ammonia is

$$H:\overset{..}{N}:H$$
$$H$$

If it grabs a nearby H atom, the NH_3 becomes NH_4 with 9 valence electrons and 4½ EPs. In this case, the NH_4 loses an electron and become NH_4^+. Since it lost an electron, you need to subtract that electron when figuring out the EPs. Now it has 8 valence electrons and 4 EPs. So its dot formula is

$$\left[\begin{array}{c} H \\ H:\overset{..}{\underset{..}{N}}:H \\ H \end{array} \right]^+$$

Notice that because NH_4^+ lost an electron, the dot formula you get now has the brackets with a plus charge outside them.

That situation, where a non-bonding EP becomes a bonding EP when it grabs a nearby atom, happens very often in chemistry. When it does, the new bond is called a **coordinate covalent bond** (ccb).

Electron Pair Geometry

The dot structures lead to a way of predicting the shapes of molecules. It is based on the **valence shell electron pair repulsion** (VSEPR) theory. That theory says that the shapes of molecules are the result of the repulsions between the electron pairs around the central atom.

Because like charges repel, the negatively charged electron pairs will get as far away from each other as possible. From the electron distribution that results, you can guess at the **electron pair geometry**; and from the electron pair geometry, you can guess at the shape of the molecule.

Most molecules have four EPs surrounding the central atom. When the four EPs repel each other, you get a **tetrahedral** electron pair geometry.

You know that molecules can have one or two double bonds. With two double bonds, like carbon dioxide (CO_2), you have two areas with electrons. When they repel each other, you get a **linear** electron pair geometry.

In the case of one double bond and two single bonds, like in formaldehyde (CH_2O), you have three areas with electrons. When they repel each other, you get a **planar trigonal** geometry.

The most common electron pair geometries are shown in Figure 4-2.

Linear   Planar Trigonal   Tetrahedral

Figure 4-2: Common Electron Pair Geometries

Shapes of Molecules

The positively charged nucleus of each atom in a molecule is not repelled by the positively charged nucleus of any other atom because they are shielded from each other by their electrons. That means they can't affect the shape of the molecule, and the molecule takes on the shape of the electron pair geometry. Depending on the electron pair geometry of the molecule and how many atoms it has, you get different **molecular shapes**.

If the molecule has a central atom with four surrounding atoms, like carbon tetrachloride (CCl_4), the four Cl atoms form a tetrahedron. So this molecular shape is called **tetrahedral**.

If the molecule has a central atom with three surrounding atoms, like ammonia (NH_3), the three H atoms form a shallow pyramid. So this molecular shape is called, you guessed it, **pyramidal**.

If the molecule has a central atom with two surrounding atoms, like carbon dioxide (CO_2), the two O atoms form a linear structure. This molecular shape is obviously called **linear**.

One exception to the linear structure is water (H_2O). For reasons that are in the Nice-to-Know area, the atoms bend into an almost 90° angle in H_2O and a few other compounds. As you might expect, this shape is called **bent**.

What is a Chemical Reaction?

The most common molecular shapes are shown in Figure 4-3.

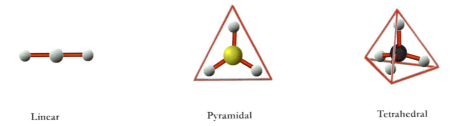

Linear   Pyramidal   Tetrahedral

Figure 4-3: Common Molecular Shapes

Now let's learn the names of the different compounds.

Homework Problems

1. Which of the following have covalent bonds?
   a. CO
   b. Fe_2O_3
   c. KCl
   d. NO
   e. None of the above

2. Which of the following have covalent bonds?
   a. SO_2
   b. Al_2O_3
   c. CCl_4
   d. FeO
   e. None of the above

3. Which of the following have ionic bonds?
   a. Hg_2
   b. NaF
   c. SO
   d. LiBr
   e. None of the above

4. Which of the following have ionic bonds?
   a. ZnO
   b. NO
   c. O_2
   d. PbO
   e. None of the above

5. Which of the following have polar covalent bonds?
   a. H_2
   b. CO
   c. KCl
   d. CCl_4
   e. None of the above

What is a Chemical Reaction?

6. Which of the following have polar covalent bonds?

   a. NO

   b. Cl_2

   c. SO_2

   d. NaCl

   e. None of the above

7. Which of the following has 4 valence electrons?

   a. As

   b. O

   c. Si

   d. I

   e. None of the above

8. Which of the following has 8 valence electrons?

   a. Ti

   b. Cl

   c. Ne

   d. Zr

   e. None of the above

9. Which of the following has 7 valence electrons?

   a. Ga

   b. I

   c. Se

   d. Ge

   e. None of the above

10. Which of the following has 5 valence electrons?

    a. Si

    b. Pb

    c. Cl

    d. Sb

    e. None of the above

11. Draw the electron dot configuration for lithium.

12. Draw the electron dot configuration for helium.

13. Draw the electron dot configuration for vanadium.

14. Draw the electron dot configuration for chlorine.

15. Draw the Lewis dot structure and structural formula for water (H_2O).

16. Draw the Lewis dot structure and structural formula for lithium oxide (Li_2O).

17. Draw the Lewis dot structure and structural formula for carbon dioxide (CO_2).

18. Draw the Lewis dot structure and structural formula for sulfur dioxide (SO_2).

19. What is the molecular shape of SO_2?

    a. tetrahedral

    b. linear

    c. pyramidal

    d. bent

    e. None of the above

20. What is the molecular shape of CO_2?

    a. tetrahedral

    b. linear

    c. pyramidal

    d. bent

    e. None of the above

21. What is the molecular shape of CCl_4?

    a. tetrahedral

    b. linear

    c. pyramidal

    d. bent

    e. None of the above

What is a Chemical Reaction?

22. What is the molecular shape of CH_3?

   a. tetrahedral

   b. linear

   c. pyramidal

   d. bent

   e. None of the above

23. What is the electron pair geometry for NH_3?

   a. tetrahedral

   b. linear

   c. pyramidal

   d. bent

   e. None of the above

24. What is the electron pair geometry for SiO_2?

   a. tetrahedral

   b. linear

   c. pyramidal

   d. bent

   e. None of the above

Chapter 5

What are Different Compounds called?

You need different names for different compounds. Otherwise how could you talk about them without getting confused? Fortunately IUPAC has set rules that tell you what to call each compound. Unfortunately the rules are complicated, and they are different for each type of compound. Names of compounds that use the IUPAC rules are called **systematic names**.

Inorganic Compounds

When it comes to naming inorganic (not containing H and C) compounds there are six different types, each with its own set of rules. They are:

1. Binary molecular compounds
2. Ternary molecular compounds
3. Binary ionic compounds
4. Ternary ionic compounds
5. Binary acids
6. Ternary acids

Binary (two) and **ternary** (three) are the number of atoms in the compound. **Molecular** (covalent and polar covalent) and **ionic** are the types of compounds. Remember covalent and ionic compounds from Chapter 4?

Acids will be discussed in a lot more detail in Chapter 12. For now just think of them as compounds whose formulas start with hydrogen (H), like HCl. **Ternary acids** always have oxygen as the third atom, like HNO_3. That's why ternary acids are sometimes called **oxyacids**.

You need to be able to go both forwards and backwards with the names of compounds. Starting with a formula, you need to be able to write its name. But staring with a name, you also need to be able to write its formula.

Binary Molecular Compounds

These are probably the easiest to name, but first you need to know the Greek number prefixes. These are listed in Table 5-1.

The rules are fairly simple:

1. Put a Greek number prefix before the first element to tell how many atoms of that element there are.
2. Put a Greek number prefix before the second element to tell how many atoms of that element there are.
3. Drop the ending on the second element and replace it with **-ide**.

For example, As_2O_3 would be diarsenic trioxide. The Greek prefixes **di-** (two) and **tri-** (three) tell you how many atoms of arsenic and oxygen there are. Then the name oxygen is changed to oxide.

Table 5-1: Greek Number Prefixes

Arabic Numeral	Greek Prefix
1	mono-
2	di-
3	tri-
4	tetra-
5	penta-
6	hexa-
7	hepta-
8	octa-
9	nona- *
10	deca-

* prefix nona- is actually Latin

There is only one exception to these rules: if the first element has a mono- in front of it, you drop the mono. For example CO_2 would be monocarbon dioxide according to rules 1 to 3. But the exception lets you drop the mono- before the carbon, so you get carbon dioxide.

Be careful! You can drop mono- before the first element, but you cannot drop mono- before the second one. The exception doesn't say anything about the second element. So only drop mono- before the first element.

You might want to ask which element goes first since neither one of them is a metal. The order established by IUPAC is this:

$$C > P > N > H > S > I > Br > Cl > O > F$$

That means C goes before any other non-metal, P goes before any other non-metal except C, N goes before any other non-metal except C or P, and so on.

You can use rules 1 to 3 in reverse to go from the name back to the formula.

Ternary Molecular Compounds

There are not very many of these compounds. Cyanate, CNO, and acetate, $C_2H_3O_2$, are about the only common ones. So we will mostly ignore ternary molecular compounds.

Binary Ionic Compounds

Ionic compounds are made up of a metal combined with a non-metal. Recall that in chemical formulas the metal always comes first and the non-metal second. The rules for naming binary ionic compounds seem easy enough:

1. Write the name of the metal element first.

2. Write the name of the non-metal element second.

3. Change the ending of the second element to –ide.

Take $MgCl_2$ as an example. First you write magnesium chlorine. Then you change the name chlorine to chloride. So $MgCl_2$ is magnesium chloride.

Here comes the complicated part. With molecular compounds you used Greek prefixes to tell how many atoms of each element there were. Here you don't do that. So how do you know how many atoms of each element there are? Well the net charge on a compound has to be zero, so you can use the number of valence electrons to find the number of atoms of each one.

Take $MgCl_2$ again as an example. From the periodic table Cl is in group 17, so it has 7 valence electrons. Being a non-metal, it will steal an electron to get 8 electrons (octet rule). So Cl has a charge of -1.

Likewise Mg is in group 2, so it has two valence electrons. Being a metal, it will give away its two electrons instead of trying to steal six. So Mg has a charge of +2.

To cancel out a +2 on Mg, you need two of the Cl atoms with a -1 charge. So there has to be one Mg and two Cl atoms, and the formula is $MgCl_2$.

That part was fairly easy. But there's one problem. Some elements have two valences. Because the d orbitals get mixed in with the s and p orbitals, sometimes they have one valence and sometimes another. For example Fe can have a +2 charge or it can have a +3 charge.

How do you tell which is which? Well you need to have more information.

The **Stock system** tells you the charge on a metal ion with a Roman numeral in parenthesis, like iron (II) oxide for Fe with a +2 charge. Iron (III) oxide is Fe with a +3 charge. When writing chemical names, you must use the Stock system for all atoms that have unpredictable valences.

When you add and subtract charges like you did before, you will find that iron (II) oxide is FeO. Iron (III) oxide is more complicated. The charge on oxygen is -2, and the charge on iron (III) is +3. If you think about it for a long time you will figure out that if you take the charge on 2 iron (III) ions, it will cancel out the charge on 3 oxygen ions. So iron (III) oxide is Fe_2O_3.

But there is a shortcut that saves you all that thinking. It is called the **crossover rule**. First write iron (III) as Fe^{+3} and oxygen as O^{-2}. Then move the superscript for O (without the minus sign) to the subscript for Fe, and the superscript for Fe to the subscript for O. You get Fe_2O_3. And if you use the crossover rule in reverse, it tells you that Fe_2O_3 is iron (III) oxide.

Other elements that have unpredictable valences are shown in Figure 5-2.

Table 5-2: Some Elements with Unpredictable Valence

	1	2		3	4	5	6	7	8	9	10	11	12	13	14	15	16	17	18
1																			
2																			
3																			
4				Ti^{+3} Ti^{+4}		Cr^{+3}	Mn^{+2}	Fe^{+2} Fe^{+3}	Co^{+2} Co^{+3}	Ni^{+2}	Cu^{+1} Cu^{+2}				As^{+3} As^{+5}			
5															Sn^{+2} Sn^{+4}			
6							W^{+3}							Hg^{+1} Hg^{+2}	Pb^{+2} Pb^{+4}			
7																			

You can use rules 1 to 3 in reverse to go from the name back to the formula.

Ternary Ionic Compounds

The names for ternary ionic compounds are a little harder. They are a metal ion followed by a non-metal **polyatomic** (more than one atom) ion. In polyatomic ions, the second atom is always O (oxygen). By the way, any atom or group of atoms that has a net charge on it must be called an **ion**.

With the polyatomic ions you often have a name ending with **–ate**. That usually means its formula ends with an O_3, like CO_3 (carbonate). But there are five exceptions: sulfate (SO_4), phosphate (PO_4), chromate (CrO_4), acetate ($C_2H_3O_2$), and cyanate (CNO). Just memorize these five, and if it is not one of them, then the –ate means O_3.

Sometimes you also have a name ending with **–ite**. That means it has one less O than the –ate. Carbonate is CO_3, so carbonite is CO_2 (one less O). Among the exceptions, sulfate is SO_4, so sulfite is SO_3 (one less O). Always write the formula for the –ate first, then take away an O to get to the –ite.

If the formula has one more O than the –ate, it is called a **per-…-ate**.

For example, ClO_3 is chlorate, so ClO_4 is perchlorate (one more O).

And if the formula has one less O than the –ite, it is called a **hypo-…-ite**.

For example, ClO_2 is chlorite, so ClO is hypochlorite (one less O).

Table 5-3 summarizes these rules using the chlorine ions as an example.

Table 5-3: Prefixes and Suffixes for Chlorine Ions

Formula	Prefix	Suffix	Name
ClO_4	per-	-ate	perchlorate
ClO_3*	-	-ate	chlorate
ClO_2	-	-ite	chlorite
ClO	hypo-	-ite	hypochlorite
Cl**	-	-ide	chloride

* Note O_3 would be O_4 or O_2 or O for the exceptions.
** Name of chlorine ion is included for completeness.

So the rules for naming ternary ionic compounds are these:

1. Write the name of the metallic ion.

2. Write the name of the polyatomic ion.

3. Replace the ending of the polyatomic ion with an –ate or –ite depending on the number of O atoms it has.

4. Add a per- or hypo- prefix to the polyatomic ion if it is needed.

A few more examples: $NaClO_3$ is sodium chlorate, so $NaClO_4$ is sodium perchlorate. Likewise $NaClO_2$ is sodium chlorite, so $NaClO$ is sodium hypochlorite.

You can use rules 1 to 4 in reverse to go from the name back to the formula.

Remember the set of flash cards just inside the back cover of this book? Included in them are some common polyatomic ions. You already cut them out, right? Now start memorizing the names of the polyatomic ions.

Binary Acids

Binary acids are also easy to name:

1. Write the prefix **hydro-** before the non-metal ion.
2. Replace the ending of the non-metal ion with the suffix **–ic**.
3. Acids are always followed by the word **acid** as a warning.

For example, HF is hydrofluoric acid, and HCl is hydrochloric acid.

You can use rules 1 to 3 in reverse to go from the name back to the formula.

Ternary Acids

As mentioned before, in ternary acids the third atom is always oxygen (O). That's why they're sometimes called **oxyacids**.

The rules for naming ternary acids are these:

1. If the polyatomic ion that contains O is an –ate, change it to **–ic**.
2. If the polyatomic ion that contains O is an –ite, change it to **–ous**.
3. If there is a per- or hypo- prefix, add it along with the –ic or -ous.
4. Acids are always followed by the word **acid** as a warning.

For example, because ClO_3 is chlorate, $HClO_3$ becomes chloric acid; and because ClO_2 is chlorite, $HClO_2$ becomes chlorous acid. Likewise, because ClO_4 is perchlorate, $HClO_4$ becomes perchloric acid; and because ClO is hypochlorite, $HClO$ becomes hypochlorous acid. See how that works?

Table 5-4 summarizes these names.

You can use rules 1 to 4 in reverse to go from the name back to the formula.

Table 5-4: Names of Ternary Acids

Ions		Acid
H + Perchlorate	becomes	Perchloric Acid
H + Chlorate	becomes	Chloric Acid
H + Chlorite	becomes	Chlorous Acid
H + Hypochlorite	becomes	Hypochlorous Acid
H + Chloride*	becomes	Hydrochloric Acid

*Name of binary acid is included for completeness.

The Language of Chemistry

Some chemists call these rules for naming compounds the **language of chemistry**. You need to know that language if you want to learn chemistry.

What are the Names of Different Compounds?

Homework Problems

1. What is the systematic name for H_2O?

    a. Dihydrogen monoxygen

    b. Hydrogen oxide

    c. Hydrogen dioxide

    d. Dihydrogen monoxide

    e. None of the above

2. What is the systematic name for CO_2?

    a. Carbon dioxygen

    b. Carbon oxygen

    c. Carbon dioxide

    d. Carbon oxide

    e. None of the above

3. What is the systematic name for MgO?

    a. Magnesium monoxygen

    b. Magnesium oxide

    c. Magnesium dioxide

    d. Magnesium monoxide

    e. None of the above

4. What is the systematic name for KCl?

    a. Potassium monochloride

    b. Potassium dichloride

    c. Potassium chloride

    d. Potassium chlorine

    e. None of the above

5. What is the systematic name for SO_2?

    a. Sulfur oxygen

    b. Sulfur dioxygen

    c. Sulfur oxide

    d. Sulfur dioxide

    e. None of the above

6. What is the systematic name for H_2S?

    a. Hydrogen sulfide

    b. Disulfur hydride

    c. Dihydrogen monosulfide

    d. Hydrogen sulfur

    e. None of the above

7. What is the Stock name for Cu_2S?

    a. Copper sulfide

    b. Copper (I) sulfide

    c. Copper (II) sulfide

    d. Copper monosulfide

    e. None of the above

8. What is the Stock name for PbO_2?

    a. Lead dioxide

    b. Lead (IV) oxide

    c. Lead oxide

    d. Lead (II) oxide

    e. None of the above

9. What is the Stock name for PbO?

    a. Lead monoxide

    b. Lead (IV) oxide

    c. Lead oxide

    d. Lead (II) oxide

    e. None of the above

10. What is the Stock name for CuS?

    a. Copper sulfide

    b. Copper (I) sulfide

    c. Copper (II) sulfide

    d. Copper monosulfide

    e. None of the above

What are the Names of Different Compounds?

11. What is the systematic name for CaS?

   a. Calcium sulfide

   b. Calcium (I) sulfide

   c. Calcium (II) sulfide

   d. Calcium monosulfide

   e. None of the above

12. What is the systematic name for Li$_2$S?

   a. Lithium sulfide

   b. Lithium (II) sulfide

   c. Lithium (I) sulfide

   d. Dilithium monosulfide

   e. None of the above

13. What is the systematic name for HCl?

   a. Chlorous acid

   b. Hydrogen chloride

   c. Chloric acid

   d. Hydrochloric acid

   e. None of the above

14. What is the systematic name for HF?

   a. Hydrogen fluoride

   b. Hydrofluoric acid

   c. Fluorous acid

   d. Fluoric acid

   e. None of the above

15. What is the systematic name for HBr?

   a. Hydrogen bromide

   b. Bromous acid

   c. Hydrobromic acid

   d. Bromic acid

   e. None of the above

16. What is the systematic name for HI?

    a. Hydroiodic acid

    b. Iodidic acid

    c. Iodidous acid

    d. Hydrogen iodide

    e. None of the above

17. What is the systematic name for KNO_2?

    a. Potassium pernitrate

    b. Potassium nitrate

    c. Potassium nitrite

    d. Potassium hyponitrite

    e. None of the above

18. What is the systematic name for Na_2SO_4?

    a. Sodium persulfate

    b. Sodium sulfate

    c. Sodium sulfite

    d. Sodium hyposulfite

    e. None of the above

19. What is the systematic name for $CaSO_2$?

    a. Calcium persulfate

    b. Calcium sulfate

    c. Calcium sulfite

    d. Calcium hyposulfite

    e. None of the above

20. What is the systematic name for MgNO?

    a. Magnesium pernitrate

    b. Magnesium nitrate

    c. Magnesium nitrite

    d. Magnesium hyponitrite

    e. None of the above

What are the Names of Different Compounds?

21. What is the systematic name for HNO_3?
    a. Pernitric acid
    b. Nitric acid
    c. Nitrous acid
    d. Hyponitrous acid
    e. None of the above

22. What is the systematic name for H_2SO_4?
    a. Sulfurous acid
    b. Hyposulfurous acid
    c. Sulfuric acid
    d. Persulfuric acid
    e. None of the above

23. What is the systematic name for H_3PO_3?
    a. Hypophosphorous acid
    b. Phosphorous acid
    c. Phosphoric acid
    d. Perphosphoric acid
    e. None of the above

24. What is the systematic name for $HClO_4$?
    a. Hypochlorous acid
    b. Chlorous acid
    c. Chloric acid
    d. Perchloric acid
    e. None of the above

Chapter 6

Credit: http://www.chemistryhelp.net/files/Balancing.png

What are Chemical Equations?

In the previous chapter you learned how to write chemical formulas. You learned, for example, that the formula for sodium chloride is NaCl. Now you need to learn how to use those formulas to write chemical equations.

Chemical Equations

A **chemical equation** is kind of like an equation in algebra, but instead of using x and y to stand for numbers, you use chemical formulas to stand for the molecules in a chemical reaction. For example, the equation for combining sodium with chlorine to make sodium chloride looks like this

$$Na + Cl_2 \rightarrow NaCl$$

Here the plus sign means "combined with", and the arrow means "makes". So this chemical equation says that sodium combined with chlorine makes sodium chloride.

A chemical formula on the left side of the arrow is the molecule you started with. It is called a **reactant**. A chemical formula on the right side of the arrow is what the reactant made. It is called a **product**. There can be more than one reactant and more than one product in a chemical equation. In the chemical equation above there are two reactants and one product.

Notice that the arrow goes from the left side to the right side. That means that the reactants make the products.

By the way, do you know what sodium chloride is? Ordinary table salt!

Balancing Chemical Equations

Notice that because it exists in nature as a diatomic molecule, the NaCl equation writes chlorine as Cl_2. But that means there are 2 chlorine atoms on the left side and only 1 chlorine atom on the right side. That can't be right! Matter is neither created nor destroyed in a chemical reaction.

So this chemical equation is not balanced. To be **balanced**, it must have the same number of atoms on the right side that it does on the left side. To balance it you need to put a 2 in front of the sodium chloride.

$$Na + Cl_2 \rightarrow 2\,NaCl$$

The 2 in front of the NaCl means there are 2 sodium chloride molecules. In other words, it is shorthand for writing NaCl + NaCl.

So you now have 2 chlorine atoms (in separate molecules) on the right side.

But you now also have 2 sodium atoms on the right side and only 1 on the left. You can fix that by putting a 2 in front of the sodium atom on the left.

$$2\,Na + Cl_2 \rightarrow 2\,NaCl$$

So you now have 2 sodium atoms and 2 chlorine atoms on the left side and 2 sodium atoms and 2 chlorine atoms on the right side. This chemical equation is now balanced, and the procedure we used to get there is called **balancing** a chemical equation.

What are Chemical Equations?

Other Symbols

Besides the symbols used above, there are others that can be used in a chemical equation. For example, if you heat the reactants to start a reaction, you put a Δ (delta) above the arrow. And if you use a catalyst like Pt (platinum) to speed up a reaction, you put a Pt above the arrow. A **catalyst** is something that speeds up a chemical reaction without taking part in it.

There is also the symbol **NR** that means no reaction occurred (happened). You should also use the symbols (s), (l), (g), or (aq) to tell whether the molecule is a solid, liquid, gas, or aqueous solution. **Aqueous solution** means that the molecule is dissolved in water.

So the NaCl equation should have been written like this

$$2\,Na\,(s) + Cl_2\,(g) \xrightarrow{\Delta} 2\,NaCl\,(s)$$

You might also write another chemical equation like this one

$$Na\,(s) + Mg\,(s) \xrightarrow{Cd} NR$$

Here Cd (cadmium) is used as a catalyst. But two metals never react with each other, so the result is NR (no reaction).

Table 6-1 lists all these symbols and their meanings.

Table 6-1: Symbols used in Chemical Reactions

Symbol	Meaning
→	makes, produces
+	combined with, added to
NR	no reaction
$\xrightarrow{\Delta}$	heat added
$\xrightarrow{Pt}$	Pt used as a catalyst
(s)	solid
(l)	liquid
(g)	gas
(aq)	aqueous solution

More Chemical Equations

Here is a reaction that has more than one molecule on each side

$$Ca\,(s) + HOH\,(l) \rightarrow Ca(OH)_2\,(aq) + H_2\,(g)$$

The 2 outside the parenthesis in $Ca(OH)_2$ means that there are two OH's in that molecule. In other words, it is shorthand for writing CaOHOH.

If you balance this equation, you get

$$Ca\ (s) + 2\ HOH\ (l) \rightarrow Ca(OH)_2\ (aq) + H_2\ (g)$$

This equation tells you what happens if you drop calcium into water. It makes aqueous (dissolved) calcium hydroxide and hydrogen gas bubbles.

Take a look at another chemical reaction

$$NaHCO_3\ (s) \xrightarrow{\Delta} Na_2CO_3\ (s) + CO_2\ (g) + H_2O\ (g)$$

If you heat sodium bicarbonate (also called sodium hydrogen carbonate), it **decomposes** (breaks up) into Na_2CO_3, CO_2, and H_2O.

When balancing chemical equations you should always start with the ugliest (most complex) molecule. In this case it is $NaHCO_3$. If you start with that and carefully count the atoms on both sides, you get this balanced equation

$$2\ NaHCO_3\ (s) \xrightarrow{\Delta} Na_2CO_3\ (s) + CO_2\ (g) + H_2O\ (g)$$

Do you know what sodium bicarbonate is? Baking soda! This equation tells you what happens to baking soda when you bake a cake. Can you see how the CO_2 and H_2O gases on the right would make the cake dough rise?

From now on you should write H_2O (water) as HOH. Do you see that both of these chemical formulas have 2 hydrogen and 1 oxygen atoms? So they are the same thing, just written differently. Using HOH for water will help you make fewer mistakes when balancing chemical equations.

Now look at this equation:

$$HNO_3\ (aq) + NH_4OH\ (aq) \rightarrow NH_4NO_3\ (aq) + HOH\ (l)$$

Count the atoms on both sides. This equation is balanced just as is.

Notice how the NO_3, NH_4 and OH sort of hang together during this reaction. They are NO_3, NH_4 and OH before the reaction, and they are still NO_3, NH_4 and OH afterwards. They do not always hang together like that, but they usually do. By the way, can you see why you should write water as HOH?

Many other polyatomic ions usually hang together in a chemical reaction. Table 6-2 shows you some of them.

You will learn more about chemical reactions in the next chapter. For now it is enough that you know how to read and write their chemical equations.

Table 6-2: Common Polyatomic Ions

Formula	Name
$C_2H_3O_2^-$	acetate ion
ClO^-	hypoclorite ion
ClO_2^-	chlorite ion
ClO_3^-	chlorate ion
ClO_4^-	perchlorate ion
CN^-	cyanide ion
CNO^-	cyanate ion
CO_2^{-2}	carbonite ion
CO_3^{-2}	carbonate ion
CrO_4^{-2}	chromate ion
$Cr_2O_7^{-2}$	dichromate ion
MnO_3^-	manganate ion
MnO_4^-	permanganate ion
NH_4^+	ammonium ion
NO_2^-	nitrite ion
NO_3^-	nitrate ion
OH^-	hydroxide ion
PO_3^{-3}	phosphite ion
PO_4^{-3}	phosphate ion
SO_3^{-2}	sulfite ion
SO_4^{-2}	sulfate ion
HCO_3^-	bicarbonate ion
HSO_4^-	bisulfate ion

Homework Problems

1. Write and balance this chemical equation: zinc burns in oxygen to form zinc oxide.

2. Write and balance this chemical equation: sodium burns in oxygen to form sodium oxide.

3. Write and balance this chemical equation: cadmium is used as a catalyst to convert carbon monoxide into carbon dioxide.

4. Write and balance this chemical equation: cadmium is used as a catalyst to convert nitrogen monoxide into nitrogen dioxide.

5. Write and balance this chemical equation: magnesium reacts with copper (II) oxide to form magnesium oxide and copper.

6. Write and balance this chemical equation: aluminum reacts with cobalt (III) oxide to form aluminum oxide and cobalt.

7. Write and balance this chemical equation: iron reacts with lead (IV) oxide to form iron (II) oxide and lead.

8. Write and balance this chemical equation: tin reacts with lead (II) oxide to form tin (IV) oxide and lead.

9. Write and balance this chemical equation: hydrochloric acid reacts with potassium hydroxide to form potassium chloride and water.

10. Write and balance this chemical equation: hydrofluoric acid reacts with sodium hydroxide to form sodium fluoride and water.

11. Write and balance this chemical equation: hydrobromic acid reacts with ammonium hydroxide to form ammonium bromide and water.

12. Write and balance this chemical equation: hydroiodic acid reacts with sodium hydroxide to form sodium iodide and water.

13. Write and balance this chemical equation: when heated, sodium bicarbonate decomposes into sodium carbonate, water, and carbon dioxide.

14. Write and balance this chemical equation: when heated, sodium bisulfate decomposes into sodium sulfate, water, and sulfur dioxide.

15. Write and balance this chemical equation: when you try to react magnesium with calcium oxide you get no reaction.

16. Write and balance this chemical equation: when you try to react aluminum with barium chloride you get no reaction.

17. Write and balance this chemical equation: potassium sulfide reacts with copper sulfate to give potassium sulfate and copper sulfide.

18. Write and balance this chemical equation: sodium sulfide reacts with zinc sulfate to give sodium sulfate and zinc sulfide.

19. Write and balance this chemical equation: magnesium nitrate reacts with water to give nitric acid and magnesium hydroxide.

20. Write and balance this chemical equation: zinc nitrate reacts with water to give nitric acid and zinc hydroxide.

21. Write and balance this chemical equation: nitric acid reacts with potassium hydroxide to give potassium nitrate and water.

22. Write and balance this chemical equation: nitric acid reacts with sodium hydroxide to give sodium nitrate and water.

23. Write and balance this chemical equation: sulfuric acid reacts with sodium hydroxide to give sodium sulfate and water.

24. Write and balance this chemical equation: sulfuric acid reacts with potassium hydroxide to give potassium sulfate and water.

Chapter 7

Credit: http://itmakessenseblog.com/files/2011/03/explosion.jpg

What is a Chemical Reaction?

You are finally ready to do some real chemistry! You needed to learn a lot of basics first, but you are now ready to see some real chemical reactions. Let's start by seeing how you know that a chemical reaction has occurred.

Signs of Chemical Reactions

There are four signs that tell you a chemical reaction has happened:

1. A change in temperature
2. A change in color
3. Forming a gas
4. Forming a precipitate

If the test tube suddenly gets hotter or colder, that is a good sign that a chemical reaction has happened. Sometimes the test tube gets hot enough to glow or even catch fire. The test tube can also get cold enough to give you frostbite. But usually it just feels a little warmer or colder in your hands.

Figure 7-1: Sodium Burning in Water

If you mix two chemicals together and they change from clear to a deep blue color, that's also a good sign that a chemical reaction has occurred. But the color change has to be permanent. Often when you mix two chemicals, the color changes for a split second while they mix, but then it gets clear again. That's not a chemical reaction. It has to change color and stay that way.

Figure 7-2: Mixing two Clear Liquids

What are Chemical Reactions?

If a gas is given off, that's another good sign that a chemical reaction has happened. The gas release can be very subtle, like a little fizzing at the bottom of a beaker. But it's usually bubbles rising through a liquid.

Figure 7-3: A Metal Reacting with Water

A **precipitate** is a substance that does not dissolve in water and settles to the bottom of the test tube. If you mix two chemicals together and something that looks like snow appears and settles very slowly to the bottom of the test tube, you're seeing a precipitate! It's another good sign that a chemical reaction has occurred.

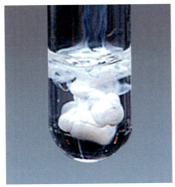

Figure 7-4: Precipitate Settling in a Test Tube

You should understand that any one of these signs by itself does not necessarily indicate that a chemical reaction has happened. For example, boiling water has bubbles rising through a liquid, but you know that boiling is a not a chemical change. But if you see two or more of these signs at the same time, you can be fairly sure that a chemical reaction has occurred.

Types of Chemical Reactions

Scientists always want to categorize things. Chemists are no exception. They divide chemical reactions into five different types:

1. Composition reactions

2. Decomposition reactions

3. Single Replacement reactions

4. Double Replacement reactions
5. Neutralization reactions

Let's take a look at what they mean by these names.

Composition Reactions

A **composition reaction** happens when two elements combine to form a compound. An example is letting iron (Fe) react with air (O_2) to form rust (FeO). Here's another example of a composition reaction:

$$C\ (s) + O_2\ (g) \xrightarrow{\Delta} CO_2\ (g)$$

In this case, carbon is burned in oxygen to make carbon dioxide.

Decomposition Reactions

A **decomposition reaction** is just the opposite of a composition reaction. It takes a compound and breaks it back down into the elements it came from.

You usually have to add heat to cause a decomposition reaction. As an example:

$$2\ HgO\ (s) \xrightarrow{\Delta} 2\ Hg\ (l) + O_2\ (g)$$

If you heat mercury (II) oxide, it turns back into mercury and oxygen.

Single Replacement Reactions

In a **single replacement reaction** one metal replaces another in a metal compound. For example:

$$2\ Al\ (s) + 3\ CuCl_2\ (aq) \rightarrow 2\ AlCl_3\ (aq) + 3\ Cu\ (s)$$

Here the Al replaces the Cu in the $CuCl_2$ and the Cu precipitates out. This is a reaction that is sometimes used to get copper metal out of copper ore.

This situation can be compared to dancing with your girl friend. Then a football player comes up and taps you on the shoulder. He winds up dancing with your girl friend, and you are left there all alone. You are the Cu.

But if you try the reaction in reverse:

$$Cu\ (s) + AlCl_3\ (aq) \rightarrow NR\ (no\ reaction)$$

As you probably guessed, the reaction goes in only one direction. Otherwise the Cu will not precipitate out like it does. So how do you know which metals replace which? The answer is in the **Activity Series**:

Table 7-5: The Activity Series

Li>K>Ba>Sr>Ca>Na>Mg>Al>Mn>Zn>Cr>Fe>Cd>Co>Ni>Sn>Pb>H>Cu>As>Ag>Hg>Au

Here Li will replace all other metals; K will replace all other metals except Li; Ba will replace all other metals except Li and K; and so on. Notice that the activity series is in the same order as increasing electro-negativity (EN). Remember EN? Because elements with lower EN lose electrons more easily than those with higher EN, they will replace them in metal compounds.

In the activity series Al is higher on the list than Cu. So the activity series tells you that Al will replace Cu, but Cu will not replace Al.

Double Replacement Reactions

In a **double replacement reaction** two metals exchange anions (partners). As an example:

$$CuCO_3 \text{ (s)} + 2 HCl \text{ (aq)} \rightarrow CuCl_2 \text{ (aq)} + H_2CO_3 \text{ (aq)}$$

Can you see how the Cu and the H trade their CO_3 and Cl anions?

Using the dance example again, there are two couples dancing, but each girl would rather dance with the other guy, so the girls swap partners. That's just what happens in a double replacement reaction.

You cannot use the activity series here because the relative EN of the metals depends on who their partner is. To try to predict which reactions will go and which will not, you need to use the **Solubility Rules**. If a reaction gives a precipitate, it is one of the signs that a reaction has occurred. The solubility rules tell you which products are insoluble and will precipitate out. So using the solubility rules, you can predict whether a reaction will happen.

Table 7-6: The Solubility Rules

Compounds that contain one of these ions are <u>soluble</u> in water:
1. alkali metal ions and the ammonium ion, Li^+, Na^+, K^+, NH_4^+
2. acetate ion, $C_2H_3O_2^-$
3. nitrate ion, NO_3^-
4. halide ions, Cl^-, Br^-, or I^- (but AgX, Hg_2X_2, & PbX_2 are insoluble)
5. sulfate ion, SO_4^{2-} (but $SrSO_4$, $BaSO_4$, & $PbSO_4$ are insoluble)
Compounds that contain one of these ions are <u>insoluble</u> in water:
6. carbonate ion, CO_3^{2-} (but rule #1 ions are soluble)
7. chromate ion, CrO_4^{2-} (but rule #1 ions are soluble)
8. phosphate ion, PO_4^{3-} (but rule #1 ions are soluble)
9. sulfide ion, S^{2-} (but CaS, SrS, BaS, & rule #1 ions are soluble)
10. hydroxide ion, OH^- (but $Ca(OH)_2$, $Sr(OH)_2$, & $Ba(OH)_2$ are soluble)

These solubility rules are the result of many years of experience by many chemists. So they are fairly reliable in predicting solubility.

Let's try out the solubility rules a little. Take this chemical reaction:

$$KNO_3 \text{ (aq)} + NaCl \text{ (aq)} \rightarrow KCl \text{ (aq)} + NaNO_3 \text{ (aq)}$$

According to solubility rule #1, KCl and $NaNO_3$ are **soluble** (dissolve in water). There is no precipitate, so the reaction does not occur. That means:

$$KNO_3 \text{ (aq)} + NaCl \text{ (aq)} \rightarrow NR \text{ (no reaction)}$$

And in fact, if you go in the laboratory and try it, you will see none of the four signs of a reaction. So the reaction does not happen.

Now look at this chemical reaction:

$$K_2S \text{ (aq)} + CuSO_4 \text{ (aq)} \rightarrow K_2SO_4 \text{ (aq)} + CuS \text{ (s)}$$

According to rule #9 of the solubility rules, CuS is **insoluble** (does not dissolve in water). So CuS is a precipitate, and the reaction does occur.

And in fact, if you go into the laboratory and try it, you will see a white precipitate. That's evidence that a chemical reaction occurred.

You should know that the solubility rules only make predictions. It is possible that a reaction the solubility rules predict will not happen, does in fact happen if you try it in the laboratory. It is also possible that a reaction the solubility rules predict will happen, does not happen in the laboratory. But predictions from the solubility rules are usually right.

Neutralization Reactions

A **neutralization reaction** is a special case of a double replacement reaction in which an acid reacts with a base. You will study acids and bases in more detail in Chapter 10. For now it is enough that you know an **acid** usually starts with H (hydrogen) and a **base** usually ends with OH (hydroxide).

The result of a neutralization reaction is a **salt** (an ionic compound) and water. Here is an example:

$$KOH \text{ (aq)} + HCl \text{ (aq)} \rightarrow KCl \text{ (aq)} + HOH \text{ (l)}$$

Here the K and H swap their partners OH and Cl, just like in a double replacement reaction. What makes it a neutralization reaction instead of a double replacement is that it happens between an acid and a base.

Notice that there is no precipitate, so the solubility rules would predict that this reaction will not happen. But if you try it in the laboratory, you will see that the test tube gets very hot. The change in temperature tells you that a chemical reaction occurred. Acids and bases always react with each other.

What are Chemical Reactions?

Homework Problems

1. Which of the following are chemical reactions? Why?
   a. Heat radiates from a burning fireplace
   b. Dry ice makes fog at a loud rock concert
   c. A rock settles to the bottom of a pan of boiling water
   d. It starts to rain during a thunder storm
   e. None of the above

2. Which of the following are chemical reactions? Why?
   a. Water drips from a closed faucet
   b. Fireworks burst across the night sky
   c. You add cream to your coffee
   d. A car comes to a stop at a stop sign
   e. None of the above

3. Which of the following are chemical reactions? Why?
   a. Water falls over a thundering waterfall
   b. A car's engine will not start
   c. A tablet of alka seltzer bubbles in a glass of water
   d. A flower opens its petals in the morning sun
   e. None of the above

4. Which of the following are chemical reactions? Why?
   a. The sky turns bright red at sunset
   b. An ice maker makes small ice cubes
   c. You add green grenadine to a glass of vodka
   d. Cake dough rises slowly in a warm pan
   e. None of the above

5. Which of the following are chemical reactions? Why?
   a. Lightning lights up the night sky
   b. Ice cubes melt in a glass of water
   c. Copper turns green after it is exposed to rain
   d. The wind picks up just before a storm
   e. None of the above

6. Which of the following are chemical reactions? Why?

   a. Dew forms on the grass on a cool morning

   b. A piece of steel wool turns reddish in moist air

   c. A rock falls down the side of a mountain

   d. Ice melts in the warm morning sun

   e. None of the above

7. Which of the following are chemical reactions? Why?

   a. A glass of Na_2CO_3 turns milky when you add calcium

   b. You are sweating in the hot sun

   c. A ball bounces off the side of a block fence

   d. Ice cream melts on a hot afternoon

   e. None of the above

8. Which of the following are chemical reactions? Why?

   a. After painting it, a wall turns bright yellow

   b. A car squeals its brakes and comes to a stop

   c. A glass of K_3PO_4 turns milky when you add lithium

   d. A ball hits a window and breaks the glass

   e. None of the above

9. Name the type of reaction: $2\ Ca + O_2 \xrightarrow{\Delta} 2\ CaO$

   a. Neutralization

   b. Single replacement

   c. Double replacement

   d. Combination

   e. None of the above

10. Name the type of reaction: $2\ K + S \xrightarrow{\Delta} K_2S$

    a. Neutralization

    b. Single replacement

    c. Double replacement

    d. Combination

    e. None of the above

What are Chemical Reactions?

11. Name the type of reaction: $Ba + MgCl_2 \rightarrow BaCl_2 + Mg$

    a. Neutralization

    b. Single replacement

    c. Double replacement

    d. Combination

    e. None of the above

12. Name the type of reaction: $2\ Al + 3\ FeO \rightarrow Al_2O_3 + 3\ Fe$

    a. Neutralization

    b. Single replacement

    c. Double replacement

    d. Combination

    e. None of the above

13. Name the type of reaction: $Na_2S + CuSO_4 \rightarrow Na_2SO_4 + CuS$

    a. Neutralization

    b. Single replacement

    c. Double replacement

    d. Combination

    e. None of the above

14. Name the type of reaction: $K_2CO_3 + MgSO_4 \rightarrow K_2SO_4 + MgCO_3$

    a. Neutralization

    b. Single replacement

    c. Double replacement

    d. Combination

    e. None of the above

15. Name the type of reaction: $KOH + HCl \rightarrow KCl + HOH$

    a. Neutralization

    b. Decomposition

    c. Double replacement

    d. Combination

    e. None of the above

16. Name the type of reaction: LiOH + HCl → LiCl + HOH

    a. Neutralization

    b. Decomposition

    c. Double replacement

    d. Combination

    e. None of the above

17. Name the type of reaction: ZnS $\xrightarrow{\Delta}$ Zn + S

    a. Neutralization

    b. Decomposition

    c. Double replacement

    d. Combination

    e. None of the above

18. Name the type of reaction: 2 CuO $\xrightarrow{\Delta}$ 2 Cu + O_2

    a. Neutralization

    b. Decomposition

    c. Double replacement

    d. Combination

    e. None of the above

19. Predict which of these reactions will occur:

    a. Mg + 2 NaOH → $Mg(OH)_2$ + 2 Na

    b. Fe + 2 NaCl → $FeCl_2$ + 2 Na

    c. 2 Al + 3 FeO → Al_2O_3 + 3 Fe

    d. Ni + ZnO → NiO + Zn

    e. None of the above

20. Predict which of these reactions will occur:

    a. 2 K + $Ba(OH)_2$ → 2 KOH + Ba

    b. Fe + BaS → FeS + Ba

    c. Al + 3 KOH → $Al(OH)_3$ + 3 K

    d. Pb + NiO → PbO + Ni

    e. None of the above

What are Chemical Reactions?

21. Predict which of these reactions will occur:
    a. $Na_2S + CaCl_2 \rightarrow CaS + 2\,NaCl$
    b. $2\,KCl + Ba(OH)_2 \rightarrow BaCl_2 + 2\,KOH$
    c. $PbS + BaCl_2 \rightarrow BaS + PbCl_2$
    d. $Na_2S + CaCl_2 \rightarrow CaS + 2\,NaCl$
    e. None of the above

22. Predict which of these reactions will occur:
    a. $Na_2S + CaCO_3 \rightarrow CaS + Na_2CO_3$
    b. $2\,SrCl_2 + Ba(OH)_2 \rightarrow BaCl_2 + Sr(OH)_2$
    c. $K_2S + BaCO_3 \rightarrow BaS + K_2CO_3$
    d. $Ag_2S + CaCl_2 \rightarrow CaS + 2\,AgCl$
    e. None of the above

23. Predict which of these reactions will occur:
    a. $Na_2S + Ca(C_2H_3O_2)_2 \rightarrow CaS + NaC_2H_3O_2$
    b. $2\,NaCl + Ba(OH)_2 \rightarrow BaCl_2 + NaOH$
    c. $Pb(NO_3)_2 + Na_2SO_4 \rightarrow 2\,NaNO_3 + PbSO_4$
    d. $Na_2S + 2\,KNO_3 \rightarrow K_2S + 2\,NaNO_3$
    e. None of the above

24. Predict which of these reactions will occur:
    a. $3\,NaC_2H_3O_2 + K_3PO_4 \rightarrow 3\,KC_2H_3O_2 + Na_3PO_4$
    b. $2\,NaCl + Li_2CrO_4 \rightarrow 2\,LiCl + Na_2CrO_4$
    c. $BaS + CaSO_4 \rightarrow CaS + BaSO_4$
    d. $3\,NH_4OH + K_3PO_4 \rightarrow 3\,KOH + (NH_4)_3PO_4$
    e. None of the above

Chapter 8

Credit: http://3.bp.blogspot.com/_AAvC0ZSs-1k/SMMCY6We3-I/AAAAAAAAy4/Z4FXaBK4qnQ/s400/Whac-a-Monty-Mole.jpg

What is a Mole?

No, it is not a small, mouse-like creature! To chemists and other scientists, a mole is a unit of measure. Just like liters tell you how much of a liquid you have, moles tell you how much of a substance you have.

When you are in a foreign country, you always have to convert from the local currency into dollars in order to understand how much something costs. That's because you usually deal in dollars, not pesos or euros or yens.

Likewise, if you tell a chemist that you have 22 liters of nitrogen or 414 grams of lead or 6 billion molecules of oil, the first thing he wants to know is "How many moles is that?" That's because chemists deal in moles, not liters or grams or molecules. So you need to know how to convert liters, grams, and molecules into moles to be able to talk to other chemists.

The Mole Thing

A **mole** is 6.02×10^{23} particles of any substance. That is an incredibly huge number! It is a 6 followed by 23 zeros. If you had that much money you could buy the entire world and still have money left over! That's how many particles there are in a mole.

The number 6.02×10^{23} is called **Avogadro's number**. How Avogadro came up with that number is in the Nice-to-Know category, so we will just skip it. Instead we will learn how you can convert a certain number of **particles** (atoms, molecules, or formula units) into moles.

Conversion Factors

To convert particles into moles, you first have to learn about conversion factors. **A conversion factor** is a ratio that allows you to convert from one set of units to another. It is similar to an exchange rate for converting pesos into dollars or dollars into pesos.

If you write the statement

$$1 \text{ dollar} = 12 \text{ pesos}$$

and divide both sides by 12 pesos, you get

$$\frac{1 \text{ dollar}}{12 \text{ pesos}} = 1$$

because 12 pesos divided by 12 pesos is equal to 1. When you do that, you change the original statement into a conversion factor that can be used to convert pesos into dollars. For example, if you are given 6 pesos in change, you can convert the pesos into dollars by multiplying them by that conversion factor.

$$6 \text{ pesos} \times \frac{1 \text{ dollar}}{12 \text{ pesos}} = 0.50 \text{ dollars}$$

You can multiply the 6 pesos by that conversion factor without changing its value because the conversion factor is equal to 1, and any number multiplied by 1 is still the same number.

What is a Mole?

Instead of dividing both sides by 12 pesos, you could have done the opposite and divided both sides by 1 dollar to get this conversion factor

$$1 = \frac{12 \text{ pesos}}{1 \text{ dollar}}$$

This conversion factor can be used to convert dollars into pesos. For example, 2.50 dollars is

$$2.50 \text{ dollars} \times \frac{12 \text{ pesos}}{1 \text{ dollar}} = 30 \text{ pesos}$$

You need to know how to make both conversion factors out of the same statement because sometimes you need to convert pesos into dollars and sometimes you need to convert dollars into pesos.

Converting Particles into Moles

The number 6.02×10^{23} particles is called the **molar number** (MN).

Just like you did with the pesos and dollars, if you take the statement

$$1 \text{ mole} = 6.02 \times 10^{23} \text{ particles}$$

you can make a conversion factor that can be used to convert particles into moles by dividing both sides by 6.02×10^{23} particles

$$\frac{1 \text{ mole}}{6.02 \times 10^{23} \text{ particles}} = 1$$

So if you have 2.71 billion particles of CO^2, you can convert them into moles like this

$$2{,}710{,}000{,}000 \text{ particles} \times \frac{1 \text{ mole}}{6.02 \times 10^{23} \text{ particles}} = 4.50 \times 10^{-15} \text{ moles}$$

Also, just like with pesos, instead of dividing both sides by 6.02×10^{23} particles, you can divide both sides by 1 mole to get this conversion factor

$$\frac{6.02 \times 10^{23} \text{ particles}}{1 \text{ mole}} = 1$$

This conversion factor can be used to convert moles into particles.

For example, if you have 0.35 moles of H_2O, then you have

$$0.35 \text{ moles} \times \frac{6.02 \times 10^{23} \text{ particles}}{1 \text{ mole}} = 2.11 \times 10^{23} \text{ particles of } H_2O$$

See how that works?

Converting Liters into Moles

It so happens that at standard temperature and pressure (STP) one mole of any gas has a volume of 22.4 L (liters). The reason for this is Nice-to-Know information, so we will just skip it.

STP means a temperature of $0°$ C and a pressure of 1 atmosphere. If it's a gas, and if it's at STP, then this statement is true

$$1 \text{ mole} = 22.4 \text{ L}$$

The volume 22.4 L at STP is called the **molar volume** (MV) of a gas.

From that statement you can make these two conversion factors

$$1 = \frac{22.4 \text{ L}}{1 \text{ mole}} \quad \text{and} \quad 1 = \frac{1 \text{ mole}}{22.4 \text{ L}}$$

The first one converts moles to liters, and the second one converts liters to moles. So if you have 2.5 moles of steam at STP, you have:

$$2.5 \text{ moles} \times \frac{22.4 \text{ L}}{1 \text{ mole}} = 56.0 \text{ L of steam}$$

Likewise if you have 67.2 L of N_2 at STP, you have:

$$67.2 \text{ L} \times \frac{1 \text{ mole}}{22.4 \text{ L}} = 3.0 \text{ moles of N2}$$

Converting Grams into Moles

It also happens that the weight of a mole of any element is equal to its atomic mass written in grams. Once again, the reason for this is Nice-to-Know information, so we will just skip it. The grams in a mole of a substance is called the **molar mass** (MM) of that substance.

For example, if you look up the atomic mass for copper in the Periodic

Table, it is 63.55 amu. So the MM for copper is 63.55 gm (grams). For lead the atomic mass is 207.2 amu. So the MM for lead is 207.2 gm.

From the statement

$$1 \text{ mole} = 207.2 \text{ gm of Pb}$$

you can make these two conversion factors

$$1 = \frac{207.2 \text{ gm}}{1 \text{ mole}} \quad \text{and} \quad 1 = \frac{1 \text{ mole}}{207.2 \text{ gm}}$$

What is a Mole?

So if you have 414.4 gm of lead, you have

$$414.4 \text{ gm} \times \frac{1 \text{ mole}}{207.2 \text{ gm}} = 2.0 \text{ moles of Pb}$$

Likewise if you have 3.0 moles of lead, you have

$$3.0 \text{ moles} \times \frac{207.2 \text{ gm}}{1 \text{ mole}} = 621.6 \text{ gm of Pb}$$

If you have a compound instead of an element, its MM is just the sum of the MM's for the elements that are in that compound. For example, the MM of KNO3 is the MM of K plus the MM of N plus 3 times the MM of O. So

$$39.10 + 14.01 + 3(16.00) = 101.11 \text{ gm}$$

Just like you did for an element, you can use the statement

$$1 \text{ mole of KNO}_3 = 101.11 \text{ gm}$$

to make two conversion factors for KNO_3.

Table 8-1: Methods for Converting to Moles

Conversion	Name	Factor	Restrictions
molar number	MN	6.02 x 10²³ particles	None
molar mass	MM	total amu's in grams	None
molar volume	MV	22.4 liters	Must be a gas at STP

More Complicated Conversions

Sometimes you know the weight of a compound but need to know the number of particles. Or you know the volume of a gas in liters but need to know its weight in grams. To find these, first convert the known quantity into moles, and then convert the moles into whatever you need to know.

Always convert the known quantity into moles first. Then you can convert from the moles into whatever it is you need to know.

For example, if you have 56.0 L of CO_2 at STP but need to know its weight, first convert 56.0 L to moles using the MV conversion factor

$$56.0 \text{ L} \times \frac{1 \text{ mole}}{22.4 \text{ L}} = 2.5 \text{ moles of CO}_2$$

Then convert moles to grams using the MM conversion factor for CO_2

$$12.01 + 2(16.00) = 44.01 \text{ gm}$$

So   2.5 moles × $\dfrac{44.01 \text{ gm}}{1 \text{ mole}}$ = 110.03 gm of CO_2

Weight of a Molecule

Using the MN, MV, and MM you can calculate many wondrous things! If someone asked you the weight of a single molecule, you would probably say you didn't know. But actually you do!

If you take 1 mole of any compound, say NO_2, because that mole of NO_2 weighs 14.01 + 2(16.00) = 46.01 gm, and because that mole of NO_2 contains 6.02 × 10²³ molecules, you can divide the total weight of a mole of NO_2 by the total number of molecules in a mole of NO_2 and get

$$\dfrac{46.01 \text{ gm}}{6.02 \times 10^{23}} = 7.64 \times 10^{-23} \text{ gm per molecule of } NO_2$$

You could do the same thing for a particle of any substance.

Density of a Gas

Likewise, knowing that a gas, say NO, weighs 14.01 + 16.00 = 30.01 gm, and that any gas at STP has a volume of 22.4 L, you can divide the total weight of a mole of NO by the total volume of that mole of NO and get

$$\dfrac{30.01 \text{ gm}}{22.4 \text{ L}} = 1.34 \text{ gm per liter of NO}$$

You may already know that mass divided by volume is called **density** and is a known number for almost all substances. It can be used to identify a gas.

Composition Percents of a Compound

If you were making enchiladas and wanted to know how much cheese is in an enchilada, how would you do that? Well you would weigh the cheese before you started and weigh the whole enchilada after you got done. Then dividing the weight of the cheese by the weight of the whole enchilada, you would get the fraction of the enchilada that is cheese. And if you wanted it in percent, you would multiply that fraction by 100%. Right?

Using MM you can do the same thing for a compound. Suppose you wanted to know how much hydrogen there is in water. Well the MM for H_2O is 2 times the MM of H plus the MM of O or 2(1.01) + 16.00 = 18.02 gm. The MM of the H alone is just 2 times the MM of H or 2(1.01) = 2.02 gm. So

$$\dfrac{2.02 \text{ gm}}{18.02 \text{ gm}} \times 100\% = 11.2\% \text{ H}$$

For the oxygen it is

$$\frac{16.00 \text{ gm}}{18.02 \text{ gm}} \times 100\% = 88.8\% \text{ O}$$

So water is 11.2% hydrogen and 88.8% oxygen. These numbers are called the **composition percents** for water.

You could do the same thing for KNO_3. The MM of KNO_3 is the MM of K plus the MM of N plus 3 times the MM of O, or

$$39.10 + 14.01 + 3(16.00) = 101.11 \text{ gm}$$

Of that total:

$$\frac{39.10 \text{ gm}}{101.11 \text{ gm}} \times 100\% = 38.7\% \text{ is K}$$

$$\frac{14.01 \text{ gm}}{101.11 \text{ gm}} \times 100\% = 13.9\% \text{ is N}$$

$$\frac{48.00 \text{ gm}}{101.11 \text{ gm}} \times 100\% = 47.5\% \text{ is O}$$

So the composition percents for KNO_3 are 38.7% K, 13.9% N, and 47.5% O.

Hydrate Complexes

Something we forgot about when we were talking about chemical formulas is **chemical complexes**. They are written with a dot between their chemical formulas and are formed by one molecule sticking to another. It's not really a chemical bond, but it's more than just standing next to each other. The molecules just sort of stick together.

One particular type of chemical complex is called a **hydrate**. It is a complex made by several water molecules sticking to a compound.

For example, copper sulfate is a hydrate that looks like this

$$CuSO_4 \bullet 5H_2O$$

The 5 means that it can have up to 5 water molecules stuck to it.

If you heat the hydrate, the water is driven off. With the water gone, it is called **anhydrous** copper sulfate.

$$CuSO_4 \bullet 5H_2O \xrightarrow{\Delta} CuSO_4 + 5H_2O$$

Small bags of anhydrous $CuSO_4$ are used to keep materials that can be damaged by moisture, like electronics, dry. Any moisture that seeps into the package is sucked up by the $CuSO_4$. Have you seen the little bags that say "Do Not Eat" in packages of electronics? Anhydrous copper sulfate!

$CuSO_4$ by itself is white, but $CuSO_4 \bullet 5H_2O$ is blue. So anhydrous $CuSO_4$ changes color when it gets moist, as shown in Picture 8-2.

Figer 8-2 : Anhydrous $CuSO_4$ Changes Color

A common question is how much water can anhydrous $CuSO_4$ absorb? You might say that you don't know. But actually you do! You can use MM and composition percents to answer that question. The MM of H_2O is 18.02, so $5H_2O$ is 90.10. The MM of anhydrous $CuSO_4$ is 159.62. Then

$$\frac{90.10 \text{ gm}}{159.62 \text{ gm}} \times 100\% = 56.4\%$$

So anhydrous $CuSO_4$ can absorb water up to 56.4% of its own weight. See what wondrous things you can do with MM!

Empirical Formula of a Compound

You can even use MM to figure out the formula of a compound. Suppose you took 2.24 gm of iron filings, heated it red hot in air, then weighed the oxidized iron that you got. If you got 3.21 gm of iron oxide, what is the formula of the iron oxide?

By conservation of mass, the oxygen in the iron oxide has to be $3.21 - 2.24 = 0.97$ gm. Using the MM of Fe (55.85 gm), and the MM of O (16.00 gm), you can calculate the moles of Fe and the moles of O in the iron oxide.

$$2.24 \text{ gm} \times \frac{1 \text{ mole Fe}}{55.85 \text{ gm}} = 0.0401 \text{ moles of Fe}$$

$$0.97 \text{ gm} \times \frac{1 \text{ mole O}}{16.00 \text{ gm}} = 0.0606 \text{ moles of O}$$

So the ratio of moles of Fe to moles of O in iron oxide is

$$0.0401 \text{ Fe to } 0.0606 \text{ O}$$

To turn these numbers into whole numbers, divide both of them by the smaller one (0.0401). The ratio will stay the same but now looks like this

$$1.00 \text{ Fe to } 1.51 \text{ O}$$

To get rid of the 1.51 fraction, multiply both numbers by 2. The ratio still stays the same but now looks like this

$$2.00 \text{ Fe to } 3.02 \text{ O}$$

If you consider that you might have made errors in your measurements, the 3.02 is probably 3.00. So the empirical formula of the iron oxide is probably

$$Fe_2O_3$$

meaning that the iron oxide contains 2 moles of Fe and 3 moles of O.

But be careful! You only get the **empirical formula** this way. The real **molecular formula** could be some number times that ratio. For example, the empirical formula for octane is C_4H_9, but its molecular formula is C_8H_{18}.

For a molecular formula, you need more information. For example, if you somehow know the real MM of a compound, that MM divided by the MM of the empirical formula tells you what number you need to multiply the empirical formula by to get the molecular formula. If you do that for octane using the molecular and empirical formulas shown above, that number turns out to be 2. So multiplying the subscripts of C_4H_9 by 2 you get C_8H_{18}.

Homework Problems

1. About how many moles is 1.50×10^{25} molecules of CO_2?
   a. 15.0 moles
   b. 1.50 moles
   c. 25.0 moles
   d. 2.50 moles
   e. None of the above

2. About how many moles is 1.80×10^{24} molecules of H_2O?
   a. 30.0 moles
   b. 3.00 moles
   c. 1.80 moles
   d. 18.0 moles
   e. None of the above

3. About how many moles is 112 grams of CaO?
   a. 0.1 moles
   b. 1.1 moles
   c. 0.2 moles
   d. 2.0 moles
   e. None of the above

4. About how many moles is 4.4 grams of CO_2?
   a. 0.1 moles
   b. 1.0 moles
   c. 0.4 moles
   d. 4.4 moles
   e. None of the above

5. About how many moles is 1.12 liters (@STP) of O_3?
   a. 0.50 moles
   b. 0.05 moles
   c. 1.12 moles
   d. 11.2 moles
   e. None of the above

What is a Mole?

6. About how many moles is 5.60 liters (@STP) of NO_2?

   a. 0.28 moles

   b. 2.80 moles

   c. 2.50 moles

   d. 0.25 moles

   e. None of the above

7. About how many moles is 2.7 x 10^{24} atoms of Fe?

   a. 35 moles

   b. 44 moles

   c. 3.5 moles

   d. 4.5 moles

   e. None of the above

8. About how many moles is 3.90 x 10^{24} atoms of Al?

   a. 3.90 moles

   b. 39.0 moles

   c. 6.50 moles

   d. 65.0 moles

   e. None of the above

9. About how many moles is 11.2 grams of C_2H_2?

   a. 2.0 moles

   b. 4.0 moles

   . 0.2 moles

   d. 0.4 moles

   e. None of the above

10. About how many moles is 226 grams of Al_3O_2?

    a. 2.0 moles

    b. 1.0 moles

    c. 0.2 moles

    d. 0.1 moles

    e. None of the above

11. About how many moles is 660 liters (@STP) of CO_2?

    a. 1.5 moles

    b. 2.0 moles

    c. 15 moles

    d. 20 moles

    e. None of the above

12. About how many moles is 709 liters (@STP) of Cl_2?

    a. 20 moles

    b. 2.0 moles

    c. 1.0 moles

    d. 10 moles

    e. None of the above

13. About how many grams does 448 liters (@STP) of F_2 weigh?

    a. 760 grams

    b. 76 grams

    c. 380 grams

    d. 38 grams

    e. None of the above

14. About how many grams does 224 liters (@STP) of H_2 weigh?

    a. 200 grams

    b. 20 grams

    c. 100 grams

    d. 10 grams

    e. None of the above

15. About how many grams do 1.5×10^{25} atoms of N_2 weigh?

    a. 70 grams

    b. 700 grams

    c. 350 grams

    d. 35 grams

    e. None of the above

What is a Mole?

16. About how many grams do 1.5 x 10²⁵ atoms of C weigh?

    a. 60 grams

    b. 600 grams

    c. 300 grams

    d. 30 grams

    e. None of the above

17. About how many liters (@STP) does 1269 grams of I_2 occupy?

    a. 2240 liters

    b. 224 liters

    c. 112 liters

    d. 1120 liters

    e. None of the above

18. About how many liters (@STP) does 799 grams of Br_2 occupy?

    a. 2240 liters

    b. 1120 liters

    c. 112 liters

    d. 224 liters

    e. None of the above

19. How many grams does a molecule of CO_2 weigh?

    a. 4.65 x 10⁻²³ grams

    b. 73.1 x 10⁻²³ grams

    c. 46.5 x 10⁻²³ grams

    d. 7.31 x 10⁻²³ grams

    e. None of the above

20. How many grams does a molecule of NO_2 weigh?

    a. 7.64 x 10⁻²³ grams

    b. 4.99 x 10⁻²³ grams

    c. 76.4 x 10⁻²³ grams

    d. 49.9 x 10⁻²³ grams

    e. None of the above

21. What is the density of CO_2 at STP?

    a. 0.98 grams/L

    b. 1.25 grams/L

    c. 1.96 grams/L

    d. 0.63 grams/L

    e. None of the above

22. What is the density of NO_2 at STP?

    a. 2.05 grams/L

    b. 1.34 grams/L

    c. 1.03 grams/L

    d. 0.67 grams/L

    e. None of the above

23. What are the composition percents for NaCl?

    a. 70.7% Na and 29.3% Cl

    b. 29.3% Na and 70.7% Cl

    c. 60.7% Na and 39.3% Cl

    d. 39.3% Na and 60.7% Cl

    e. None of the above

24. What are the composition percents for FeO?

    a. 32.3% Fe and 67.7% O

    b. 67.7% Fe and 32.3% O

    c. 77.7% Fe and 22.3% O

    d. 22.3% Fe and 77.7% O

    e. None of the above

Chapter 9

Credit: http://www.searchamateur.com/pictures/whack-a-groundhog-1.jpg

How many Moles?

In the previous chapter you learned about moles. You saw how to convert from particles, grams, or liters to moles and vice versa. Let's find out what you can do with that!

Mole Reactions

Up to now, when you wrote this balanced chemical equation

$$CuCO_3 \text{ (s)} + 2 \text{ HCl (aq)} \rightarrow CuCl_2 \text{ (aq)} + CO_2 \text{ (g)} + HOH \text{ (l)}$$

you meant that 1 molecule of $CuCO_3$ reacts with 2 molecules of HCl to make 1 molecule of $CuCl_2$, 1 molecule of CO_2, and 1 molecule of water.

But what if you had two of these reactions going on side by side? Well you could write two copies of the equation next to each other, but that seems a bit silly. Instead, you would probably write only one equation but say that it stood for two identical reactions going on side by side.

What if you had 1000 identical reactions going on side by side? Rather than write 1000 copies of the equation next to each other, you would do the same thing you did before and write just one equation but say that it stood for 1000 identical reactions going on side by side.

Well what if you had 6.02×10^{23} identical reactions going on side by side? It would take a very long time to write 6.02×10^{23} copies of the equation next to each other. Instead you would probably do the same thing as before and write just one equation but say that it stood for 6.02×10^{23} (one mole) identical reactions going on side by side.

That leads to a new meaning for this chemical equation

$$CuCO_3 \text{ (s)} + 2 \text{ HCl (aq)} \rightarrow CuCl_2 \text{ (aq)} + CO_2 \text{ (g)} + HOH \text{ (l)}$$

It now means that 1 mole of $CuCO_3$ reacts with 2 moles of HCl to make 1 mole of $CuCl_2$, 1 mole of CO_2, and 1 mole of water.

Mole to Mole Conversions

Suppose you had 2.50 moles of HCl and wanted to know how many moles of $CuCO_3$ you needed for this reaction. Well according to the balanced chemical equation shown above, it takes 1 mole of $CuCO_3$ to react with 2 moles of HCl. So the ratio of moles of $CuCO_3$ to moles of HCL is

$$\frac{1 \text{ mole of } CuCO_3}{2 \text{ moles of HCl}}$$

If you look at it as a conversion factor (which it is), that ratio can be used to convert moles of HCl into moles of $CuCO_3$. So

$$2.50 \text{ moles of HCl} \times \frac{1 \text{ mole of } CuCO_3}{2 \text{ moles of HCl}} = 1.25 \text{ moles of } CuCO_3$$

How many Moles is that?

You could also use the inverse of that ratio

$$\frac{2 \text{ moles of HCl}}{1 \text{ mole of CuCO}_3}$$

to convert from moles of $CuCO_3$ to moles of HCl. Suppose you had 1.75 moles of $CuCO_3$ and wanted to know how many moles of HCl you needed.

$$1.75 \text{ moles of CuCO}_3 \times \frac{2 \text{ moles of HCl}}{1 \text{ mole of CuCO}_3} = 3.50 \text{ moles of HCl}$$

First you need to balance the chemical equation, but then you can use it to make several other mole-to-mole conversion factors.

$$\frac{1 \text{ mole of CuCl}_2}{2 \text{ moles of HCl}} \quad \text{and} \quad \frac{2 \text{ moles of HCl}}{1 \text{ mole of CuCl}_2}$$

$$\frac{1 \text{ mole of CO}_2}{2 \text{ moles of HCl}} \quad \text{and} \quad \frac{2 \text{ moles of HCl}}{1 \text{ mole of CO}_2}$$

$$\frac{1 \text{ mole of CuCl}_2}{1 \text{ mole of CuCO}_3} \quad \text{and} \quad \frac{1 \text{ mole of CuCO}_3}{1 \text{ mole of CuCl}_2}$$

$$\frac{1 \text{ mole of CO}_2}{1 \text{ mole of CuCO}_3} \quad \text{and} \quad \frac{1 \text{ mole of CuCO}_3}{1 \text{ mole of CO}_2}$$

These can be used to convert from moles of one of the **reactants** (compounds on the left side of the equation) to moles of one of the **products** (compounds on the right side of the equation) or vice versa.

Do another example. Suppose you want 4.15 moles of CO_2 (one of the products). How many moles of HCl (one of the reactants) do you need? If you choose the right mole-to-mole conversion factor

$$4.15 \text{ moles of CO}_2 \times \frac{2 \text{ moles of HCl}}{1 \text{ mole of CO}_2} = 8.30 \text{ moles of HCl}$$

See how that works?

More Complicated Conversions

What if instead of moles of $CuCO_3$, you were given grams of $CuCO_3$ and asked how many liters of CO_2 you would get. Well this is where you use the MM to convert grams to moles and MV to convert moles to liters.

For example, suppose you were given 61.78 grams of $CuCO_3$ and asked how many liters of CO_2 it would make.

The MM for $CuCO_3$ is 63.55 + 12.01 + 3(16.00) = 123.56 gm. So

$$61.78 \text{ gm } CuCO_3 \times \frac{1 \text{ mole } CuCO_3}{123.56 \text{ gm } CuCO_3} = 0.500 \text{ moles } CuCO_3$$

Using the right mole-to-mole ratio from the balanced equation above

$$0.500 \text{ moles } CuCO_3 \times \frac{1 \text{ mole of } CO_2}{1 \text{ mole of } CuCO_3} = 0.500 \text{ moles } CO_2$$

The MV of any gas is 22.4 L at STP, so

$$0.500 \text{ moles } CO_2 \times \frac{22.4 \text{ L } CO_2}{1 \text{ mole } CO_2} = 11.2 \text{ L } CO_2 \text{ @ STP}$$

This is what you do: first convert the mass of the $CuCO_3$ to moles using its MM. Then use the right mole-to-mole conversion factor from the balanced equation to find the moles of CO_2. Finally use the MV to find the volume of the CO_2. It seems very complicated now, but it will get easier with practice.

Take another example. Suppose someone asked you how many liters of CO_2 6,000,000,000 molecules (mlc) of HCl would make. You would probably say you didn't know. But actually you do! Using the MN

$$6.0 \times 10^9 \text{ mlc HCl} \times \frac{1 \text{ mole HCl}}{6.02 \times 10^{23} \text{ mlc HCl}} = 1.0 \times 10^{-14} \text{ moles HCl}$$

Now using the right mole-to-mole conversion factor from above

$$1.0 \times 10^{-14} \text{ moles HCl} \times \frac{1 \text{ mole } CO_2}{2 \text{ moles HCl}} = 0.5 \times 10^{-14} \text{ moles } CO_2$$

Finally using the MV of any gas at STP you get

$$0.5 \times 10^{-14} \text{ moles } CO_2 \times \frac{22.4 \text{ L } CO_2}{1 \text{ mole } CO_2} = 11.2 \times 10^{-14} \text{ L } CO_2$$

How many Moles is that?

Limiting Reactants

Up to now you assumed that you had just the right amount of each of the reactants. But what if you had more of one than you need? Well the reaction would use up all of the other reactant, and you would have some of the first one left over. The substance that is all used up is called the **limiting reactant** because it limits how much of the products you can make. The substance that is left over is called the **excess reactant**.

So is the limiting reactant $CuCO_3$ or HCl in the reaction we used before? To answer that you need to find out how much CO_2 the $CuCO_3$ would make and how much CO_2 the HCl would make. The one that would make the least amount of CO_2 is the limiting reactant and would be all used up before the other one.

Take an example. Suppose you have 617.8 gm of $CuCO_3$ and 109.4 gm (before dissolving) of HCl. If you use the MM for $CuCO_3$, 617.8 gm is 5.00 moles, and if you use the MM for HCl, 109.4 gm is 3.00 moles. So which one is the limiting reactant and which one is the excess reactant?

Using the right mole-to-mole ratios from before, for the $CuCO_3$ you get

$$5.00 \text{ moles } CuCO_3 \times \frac{1 \text{ mole of } CO_2}{1 \text{ mole of } CuCO_3} = 5.00 \text{ moles } CO_2$$

For the HCl, you get

$$3.00 \text{ moles of HCl} \times \frac{1 \text{ mole of } CO_2}{2 \text{ moles of HCl}} = 1.50 \text{ moles of } CO_2$$

Since 617.8 gm of $CuCO_3$ can make 5.0 moles of CO_2 but 109.4 gm of HCl can make only 1.5 moles of CO_2, the HCl will run out before the $CuCO_3$ does. So HCl is the limiting reactant, and $CuCO_3$ is the excess reactant.

Percent Yield

In the real world you sometimes don't get as much of one of the products as you think you should. The reason could be that you made a mistake in calculating how much of a reactant you needed. Or it could be that you made a mistake in measuring one of the reactants. It could also be that some of the reactants could not find each other in order to react completely.

Regardless of why, you get less (or possibly more) of the products than you thought you should get. That leads to the idea of **percent yield**.

$$\text{Percent Yield} = \frac{\text{Actual Yield}}{\text{Calculated Yield}} \times 100\%$$

The calculated yield is sometimes also called the **theoretical yield**.

Take an example. In the reaction you have been using, suppose you calculate that you should get 11.2 L of CO_2, but you really get only 9.8 L. The percent yield is

$$\text{Percent Yield} = \frac{9.8 \text{ L } CO_2}{11.2 \text{ L } CO_2} \times 100\% = 87.5\%$$

Sometimes you get more from a reaction than you thought you should get. In the real world you really can't get more than 100% yield from a chemical reaction. So if you see a percent yield that is more than 100%, look for a mistake in either your calculations or your measurements.

How many Moles is that?

Homework Problems

Looking at this reaction @ STP: $2H_2\ (g) + O_2\ (g) \rightarrow 2H_2O\ (g)$
1. If you use 1.00 mole of H_2, how many moles of O_2 are used up?

   a. 2.00 moles

   b. 0.50 moles

   c. 1.50 moles

   d. 1.00 moles

   e. None of the above

Looking at this reaction @ STP: $2H_2\ (g) + O_2\ (g) \rightarrow 2H_2O\ (g)$
2. If you use 1.00 mole of H_2, how many moles of H_2O are made?

   a. 2.00 moles

   b. 0.50 moles

   c. 1.50 moles

   d. 1.00 moles

   e. None of the above

Looking at this reaction @ STP: $2Fe\ (s) + 3O_2\ (g) \rightarrow 2Fe_2O_3\ (s)$
3. If you use 1.50 moles of Fe, how many moles of Fe_2O_3 are made?

   a. 1.50 moles

   b. 2.25 moles

   c. 0.75 moles

   d. 1.00 moles

   e. None of the above

Looking at this reaction @ STP: $2Fe\ (s) + 3O_2\ (g) \rightarrow 2Fe_2O_3\ (s)$
4. If you use 1.50 moles of Fe, how many moles of O_2 are used up?

   a. 1.50 moles

   b. 2.25 moles

   c. 0.75 moles

   d. 1.00 moles

   e. None of the above

Looking at this reaction @ STP: 4Na (s) + O_2 (g) → $2Na_2O$ (s)

5. If you use 3.00 moles of Na, how many moles of O_2 are used up?

   a. 1.50 moles

   b. 0.50 moles

   c. 0.75 moles

   d. 1.00 moles

   e. None of the above

Looking at this reaction @ STP: 4Na (s) + O_2 (g) → $2Na_2O$ (s)

6. If you use 3.00 moles of Na, how many moles of Na_2O are made?

   a. 1.50 moles

   b. 0.50 moles

   c. 0.75 moles

   d. 1.00 moles

   e. None of the above

Looking at this reaction @ STP: 2Fe (s) + $3O_2$ (g) → $2Fe_2O_3$ (s)

7. If you use 14 grams of Fe, how many liters of O_2 are needed?

   a. 4.2 liters

   b. 2.8 liters

   c. 5.6 liters

   d. 8.4 liters

   e. None of the above

Looking at this reaction @ STP: 2Fe (s) + $3O_2$ (g) → $2Fe_2O_3$ (s)

8. If you use up 12 liters of O_2, how many grams of Fe_2O_3 are made?

   a. 128 grams

   b. 57 grams

   c. 64 grams

   d. 114 grams

   e. None of the above

How many Moles is that?

Looking at this reaction @ STP: 4Na (s) + O_2 (g) → $2Na_2O$ (s)

9. If you use up 11 liters of O_2, how many grams of Na_2O are made?

   a. 15 grams

   b. 31 grams

   c. 61 grams

   d. 122 grams

   e. None of the above

Looking at this reaction @ STP: 4Na (s) + O_2 (g) → $2Na_2O$ (s)

10. If you use 16 grams of Na, how many liters of O_2 are needed?

    a. 7.8 liters

    b. 1.9 liters

    c. 6.2 liters

    d. 3.9 liters

    e. None of the above

Looking at this reaction @ STP: $2H_2$ (g) + O_2 (g) → $2H_2O$ (g)

11. If you use 2.8 liters of H_2, how many liters of O_2 are needed?

    a. 4.2 grams

    b. 1.4 liters

    c. 2.8 liters

    d. 7.0 liters

    e. None of the above

Looking at this reaction @ STP: $2H_2$ (g) + O_2 (g) → $2H_2O$ (g)

12. If you use 1.1 grams of H_2, how many grams of H_2O are made?

    a. 9.8 grams

    b. 4.9 grams

    c. 20 grams

    d. 3.7 grams

    e. None of the above

Looking at this reaction @ STP: 2Ca (s) + O_2 (g) → 2CaO (s)

13. If 10 gm of Ca react with 6 L of O_2, what is the limiting reactant?

    a. Ca

    b. CaO_2

    c. O_2

    d. CaO

e. None of the above

Looking at this reaction @ STP: 2Ca (s) + O_2 (g) → 2CaO (s)

14. If 7 gm of Ca react with 3 L of O_2, what is the excess reactant?

    a. CaO

    b. Ca

    c. O_2

    d. CaO_2

    e. None of the above

15. If 7.0 gm of CO react with 3.0 L of O_2, how much CO_2 do you get?

    a. 2.8 L

    b. 3.5 L

    c. 5.6 L

    d. 3.0 L

    e. None of the above

16. If 3.0 L of CO react with 4.0 gm of O_2, how much CO_2 do you get?

    a. 1.5 L

    b. 2.8 L

    c. 3.0 L

    d. 5.6 L

    e. None of the above

17. If 5.6 gm of Fe react with 2.0 L of O_2, how much Fe_2O_3 do you get?

    a. 2.0 gm

    b. 9.5 gm

    c. 8.0 gm

    d. 5.6 gm

    e. None of the above

How many Moles is that?

18. If 5.4 gm of Fe react with 1.5 L of O_2, how much Fe_2O_3 do you get?

   a. 7.1 gm

   b. 3.9 gm

   c. 3.6 gm

   d. 7.7 gm

   e. None of the above

19. You get 4.0 gm of CO_2 when you burn 1.2 gm of C. What is the yield?

   a. 89% yield

   b. 93% yield

   c. 87% yield

   d. 91% yield

   e. None of the above

20. You get 5.7 gm of SO_2 when you burn 3.2 gm of S. What is the yield?

   a. 89% yield

   b. 93% yield

   c. 87% yield

   d. 91% yield

   e. None of the above

21. You get 2.5 gm of CO when you burn 1.2 gm of C. What is the yield?

   a. 87% yield

   b. 91% yield

   c. 89% yield

   d. 93% yield

   e. None of the above

22. You get 5.2 gm of CaO when you heat 4.0 gm of Ca. What is the yield?

   a. 87% yield

   b. 91% yield

   c. 89% yield

   d. 93% yield

   e. None of the above

23. You get 4.0 gm of NO_2 when you heat 2.8 gm of N_2. What is the yield?

   a. 87% yield

   b. 91% yield

   c. 89% yield

   d. 93% yield

   e. None of the above

24. You get 1.6 gm of H_2O when you burn 0.2 gm of H_2. What is the yield?

   a. 87% yield

   b. 91% yield

   c. 89% yield

   d. 93% yield

   e. None of the above

Chapter 10

Credit: http://www.inyoprocess.com/components/com_fpss/images/gas_liquid_injector_co2.jpg

What is a Gas? A Liquid? A Solid?

In the first chapter you learned some things about solids, liquids, and gases and the differences between them. To remind you, Table 10-1 shows you that information again. But there are lots of other things about solids, liquids, and gases that are not shown in Table 10-1.

Table 10-1: The Forms of Matter

	Volume	**Shape**	**Compressible?**
Gas	Variable	Variable	Yes
Liquid	Definite	Variable	No
Solid	Definite	Definite	No

Gases

Table 10-1 tells you that a gas can be compressed (squeezed together) a lot. That means it does not have a definite volume. Why is that?

The answer is that the distance between molecules in a gas is very large compared to the distance between molecules in a liquid or a solid. In fact, that distance can be thousands of times greater, and it can be changed a lot just by increasing or decreasing the **pressure** (push per square inch) on the gas.

When you increase the pressure on a gas, the distance between the molecules gets a little smaller. And when you decrease the pressure on a gas, the distance between the molecules gets a little larger. Can you see how easy it is for a gas to compress or expand?

If you can compress a gas a lot, you should be able to squeeze it into a bottle. If you do that, the gas takes on the shape and volume of the bottle. You could also squeeze the gas into a balloon. Then it takes on the shape and volume of the balloon. So it can be like a bottle one minute and like a balloon the next! That's why we say that a gas does not have a definite volume or a definite shape.

This brings us to the idea of density. **Density** is the mass of a substance divided by its volume. Because the volume of a gas can increase so much, a gas can have a much lower density than a solid or a liquid. In fact, it can be thousands of times less dense than a solid or a liquid of the same weight.

Gases form Homogeneous Mixtures

All gases mix easily with other gases and quickly form homogeneous mixtures. Air, for example, is mostly a mixture of oxygen and nitrogen. It is about 20% oxygen and 80% nitrogen.

You can be perfectly sure that air is a homogeneous mixture because the students sleeping in the back of the room right now don't turn blue for lack of oxygen. They get the same 20% oxygen you do. In fact, so do the students in all the classrooms around the world.

The Ideal Gas Law

Lowering the pressure on a gas increases its volume. Besides pressure, there is one other thing that changes the volume of a gas. If you raise the temperature of a gas, it expands, and that increases the volume of the gas.

The magnitudes of the pressure (P), temperature (T), and volume (V) all depend on each other. Skipping all the Nice-to-Know reasons, P, V, and T are related to each other by this equation

$$P \times V = n \times k \times T$$

It is called the **ideal gas law**. Here n is the number of moles, and k is a constant number equal to 0.0821 (L · atm)/ (mol · °K). The atm stands for the pressure in atmospheres, and the mol stands for the number of moles. In this equation the temperature (T) has to be in degrees Kelvin (°C + 273).

This law is called ideal because it is for the behavior of an ideal gas. For various reasons (all of them Nice-to-Know) gases are not really ideal, but their behavior is close enough to that of an ideal gas that this equation gives fairly good results for P, V, and T.

If the number of moles n and the temperature T are forced to stay the same, then (n x k x T) is a constant number (remember that k is a constant). In that case the pressure times the volume on one day is equal to the pressure times the volume on some other day. So

$$P_2 \times V_2 = (n \times k \times T) = P_1 \times V_1$$

For example, if the pressure yesterday was 1.00 atm (P_1) for a gas with a volume of 100 L (V_1), and the pressure today is 0.98 atm (P_2), what is the volume (V_2) of that gas today? The ideal gas law says

$$V_2 = (P_1 \times V_1)/ P_2 = 102 \text{ L}$$

What if the temperature changes too? Well as long as the number of moles doesn't change, then (n x k) is a constant number, and the pressure times the volume divided by the temperature on one day is equal to the pressure times the volume divided by the temperature on some other day. In that case

$$\frac{P_2 \times V_2}{T_2} = (n \times k) = \frac{P_1 \times V_1}{T_1}$$

In this form, the ideal gas law is called the **combined gas law**.

Using the combined gas law, if the volume (V_1) of a gas is 20 L at a pressure (P_1) of 1.5 atm and a temperature (T_1) of 300 °K, what is its volume at STP? STP, if you remember, is a pressure (P_2) of 1.0 atm and a temperature (T_2) of 0 °C (273 °K). So according to the combined gas law

$$V_2 = \frac{P_1}{P_2} \times \frac{T_2}{T_1} \times V_1 = 33 \text{ L}.$$

Can you see how this equation lets you adjust the volume of a gas that is not at STP to what it would be at STP? Then you can use the MV conversion factor (1 mole = 22.4 L @ STP) to change the volume of the gas into moles.

Atmospheric Pressure

We keep using the term atmospheres to measure pressure. So what is an atmosphere? Well, gases do not weigh very much, but they do have some weight. If you stack a column of air from the earth's surface to the edge of space (about 100 km high), it will actually weigh 14.7 psi (pounds per square inch). It is this weight of air that we call atmospheric pressure, and 14.7 psi is the pressure of one atmosphere. It is sometimes measured in millimeters (mm) of mercury for reasons that are Nice-to-Know, so we skip them. It is enough that you know that 760 mm of mercury is the same as 14.7 psi, the pressure of **one atmosphere**.

Liquids

Table 10-1 says that liquids, like gases, do not have a definite shape. Liquids flow easily and take on the shape of their container. If you pour a liquid into a bottle, it takes on the shape of the bottle. If you pour it into a wine glass, it takes on the shape of the wine glass. So it does not have a definite shape.

Liquids and gases are called **fluids** because they flow so easily. They flow easily because their molecules are not held in place like they are in a solid. They can slide past each other. That's why they don't have a definite shape.

Table 10-1 also says that liquids, unlike gases, cannot be compressed very much. Because the distance between molecules in a liquid is much smaller than the distance between molecules in a gas, liquids cannot be compressed like gases can. In fact, the distance between molecules in a liquid is only slightly larger than the distance between molecules in a solid. So liquids cannot be compressed very much because their molecules are already rubbing shoulders. This means that they have a definite volume.

Intermolecular Bonds in Liquids

Up until now we have considered only **intra-molecular bonds** (the attraction between atoms in a single molecule). In liquids and solids the molecules are so close together that you need to consider **inter-molecular bonds** (the attraction between the atoms of two molecules that are close to each other). These inter-molecular bonds about 10 times weaker than intra-molecular bonds, but they can still affect the behavior of substances.

Chemists recognize three types of inter-molecular bonds (sometimes called inter-molecular forces): dispersion bonds, dipole bonds, and hydrogen bonds. **A dispersion bond** is the weakest of the inter-molecular bonds. To understand it you need to form a mental picture of a molecule. Electrons are always flying around the nucleus of a molecule at a high rate of speed. Picture them like a swarm of bees flying madly around their hive. At any moment, more of the electrons might be on the left side than on the right side, or vice versa. For a split second then, the molecule has a negative charge on one side and a positive charge on the other. The same thing happens with all the other nearby molecules. If the temporarily negative charge of one molecule is near the temporarily positive charge of a nearby molecule, the two of them attract each other. That attraction is a dispersion bond. All molecules have dispersion bonds that last only for a split second.

The second strongest inter-molecular bond is the **dipole bond**. In chapter 4 you learned about polar molecules. If you remember, they are molecules formed between two atoms that do not share their electrons equally. The electrons spend more of their time around one of the atoms than they do around the other. That causes a situation where the electrons are more on one side of the molecule than they are on the other. That is exactly what happens with a dispersion bond, but here the bond does not last only a split second. So that's just what a dipole bond is: a permanent dispersion bond.

The strongest inter-molecular bond is the **hydrogen bond**. It is a special type of dipole bond where the atoms involved are hydrogen with either oxygen, nitrogen, or fluorine. They form such an unusually strong dipole bond that they are given a separate name.

Surface Tension in Liquids

Great! There are three kinds of inter-molecular bonds. What's your point?

Well the point is that the inter-molecular bonds explain a lot about liquids. In fact, the surface tension of a liquid is higher or lower depending on the strength of the inter-molecular bonds between molecules. **Surface tension** is caused by the attraction of nearby molecules at the surface of a liquid. This inter-molecular attraction between adjacent molecules forms a kind of trampoline that keeps other molecules from breaking through the surface.

If you don't believe it, watch a wasp drinking water from your swimming pool. It lands on the surface of the pool but does not sink! In order for an object to sink it first has to break through the surface tension of the liquid. It takes a certain amount of weight to do that. Now look at what the wasp is doing. If you look closely, you will see that it has spread its legs way out. It does that to spread its weight over a larger area and reduce its weight per square inch. That lets the surface tension of the water support its weight. The wasp kind of walks on water! Now do you believe in surface tension?

Boiling Point of Liquids

The same can be said about the boiling point of a liquid. It is higher or lower depending on the strength of its intermolecular bonds. In order for a molecule to leave a liquid, it first has to break through the surface tension. That takes a certain amount of energy. The molecule gets that energy from the temperature of the liquid. As the temperate increases, the kinetic energy of the molecule increases. When it gets high enough, the molecule can break through the surface tension and leave.

Liquids with weak inter-molecular bonds can boil at a lower temperature because they have weak surface tensions. Liquids with strong inter-molecular bonds, on the other hand, require a much higher temperature to boil because they have much stronger surface tensions.

Vapor Pressure of Liquids

The vapor pressure of a liquid goes just the opposite of the boiling point. **Vapor pressure** is the pressure above a liquid caused by evaporation. If the boiling point is very high, less of the liquid evaporates at room temperature, so the vapor pressure is low. If the boiling point is low, more of the liquid evaporates at room temperature, so the vapor pressure is high.

Viscosity of Liquids

The same can be said of the viscosity of a liquid. It is higher or lower depending on the strength of its inter-molecular bonds. **Viscosity**, you probably already know, is the ability of a liquid to flow. Molasses flow very slowly because they have a high viscosity. Rubbing alcohol flows easily because it has a low viscosity. In the case of molasses, the inter-molecular bonds are strong. In the case of alcohol, the inter-molecular bonds are weak.

Did you know that glass is a liquid? It is, but it has an incredibly high viscosity because of its incredibly high inter-molecular bonds. It actually takes hundreds of years for it to flow very far. The stained glass windows in the Sistine Chapel at the Vatican were installed about 500 years ago, and they are just now beginning to show a measurable difference between their thickness at the top and their thickness at the bottom!

Solids

Table 10-1 says that solids, unlike fluids, have a definite shape. They also cannot be compressed very much, so they have a definite volume.

Substances are usually denser as solids than they are as liquids. That is because the molecules are a little closer together in solids than they are in liquids, so their volume is smaller.

Water and ammonia are two major exceptions. The hydrogen bonds in these liquids make them form a ring structure when they freeze into a solid. That makes them expand slightly, and that reduces their density. This is why ice cubes float on water. Frozen ammonia also floats on liquid ammonia.

Crystalline Forms of Solids

In most solids the molecules are arranged in a regular pattern called **crystals**. Chemists recognize three forms of crystals: ionic, molecular, and metallic. Just like with liquids, the physical properties of these crystals depend on the relative strength of their inter-molecular bonds.

Ionic Crystals

In **ionic crystals** there is a pattern of alternating positive and negative ions. NaCl (table salt) is an ionic crystal and consists of alternating Na^+ and Cl^- ions. Its structure is shown in Figure 10-2.

What is a Gas? A Liquid? A Solid?

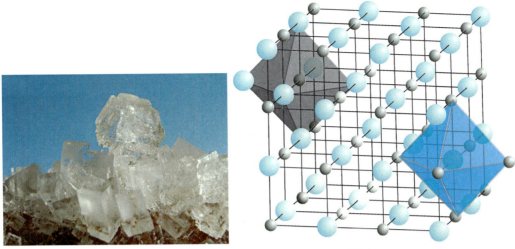

Figure 10-2: Crystalline Structure of Salt

Most ionic compounds (a metal + a non-metal) form ionic crystals. They are usually hard and **brittle** (break easily). They are also usually **soluble** (dissolve completely) in water. Because of their regular structure, light can pass between the atoms of ionic crystals. They are said to be **translucent**.

Molecular Crystals

Sugar, ($C_{12}H_{22}O_{11}$), on the other hand, is a **molecular crystal**. Its structure is shown in Figure 10-3. Here the repeating pattern is between the much weaker δ^+ and δ^- partial charges found in covalent molecules.

Figure 10-3: Crystalline Structure of Sugar

Most covalent (a non-metal + another non-metal) compounds form molecular crystals. They are usually **insoluble** (do not dissolve) in water. Some molecular crystals, like sugar, are translucent, but most are not.

Metallic Crystals

The third type of crystalline solids is **metallic crystals**. Cu (copper) is a metallic crystal. Its structure is shown in Figure 10-4. Here the atoms are stacked like cannonballs.

 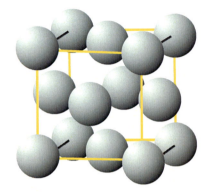

Figure 10-4: Crystalline Structure of Copper

Most metals are metallic crystals. They are **ductile** (can be pulled into wires), **malleable** (can be flattened into thin foils), and are good **conductors** of electricity (electrons flow through them easily). They are usually insoluble in water.

Melting Point of Solids

In gases the distance between molecules is so large that you can ignore inter-molecular bonds. In liquids and solids, however, the distance between molecules is much smaller, so the inter-molecular bonds very important.

You already know about the inter-molecular bonds in liquids and know that many of the physical properties of liquids can be explained by the strength of their inter-molecular bonds. The same is true for solids.

One important physical property of a solid is its melting point. It takes energy to break the inter-molecular bonds when a substance melts, and stronger inter-molecular bonds need more energy to break them apart than weaker inter-molecular bonds.

In ionic and molecular crystals the inter-molecular bonds are like the permanent dipole bonds in liquids. But obviously, the inter-molecular bonds are much greater between the full (+1 and -1) electrical charges in ionic crystals than they are between the partial (δ^+ and δ^-) electrical charges in molecular crystals. That explains why ionic crystals have a much higher melting point than molecular crystals.

Metallic crystals have inter-molecular bonds somewhere between those of ionic and molecular crystals. So they have melting points somewhere between the high ones for ionic and the low ones for molecular crystals.

Other physical properties of solids can be explained by the strength of their inter-molecular bonds. But for right now just understanding the relation between inter-molecular bonds and melting points is good enough. Other properties of the different crystalline solids are shown in Table 10.5.

Table 10-5: Properties of Crystalline Solids

	Melting Point	Soluble in Water	Hard & Brittle	Translucent	Maleable & Ductile	Electtrical Conductor
Lonic	High	Yes*	Yes*	Yes*	No**	No**
Molecular	Low	No**	No**	No**	No**	No**
Metallic	Intermediate	No**	No**	No**	Yes*	Yes*

* Usually, but there are exceptions   ** Usually not, but there are exceptions

Changes from a Solid to a Liquid to a Gas

You know that by adding heat you can change (melt) a solid into a liquid. By continuing to add heat, you can also change (boil) a liquid into a gas. But how much heat do you need?

To answer that question you first need to learn about heat capacity. The **heat capacity** of a substance tells you how much heat you need to raise 1 gm of the substance by 1 °C. It is measured in calories and is defined such that it is exactly 1.0 for (pure) water. So a **calorie** is the heat that will raise 1 gm of water by 1 °C. Heat capacity has different values for different materials and can be higher or lower than 1.0 depending on the material.

That's the heat capacity for liquid water. For ice (a solid) or steam (a gas) it is about half of that (0.5). For Nice-to-Know reasons that we will skip, this is true for almost all substances. Their heat capacity as a solid or a gas is about half of what it is as a liquid.

So how much heat is needed to raise 25.0 gm of water from 0 °C (the freezing point) to 100 °C (the boiling point)? The equation for this is

$$Q = c \times M \times \Delta T$$

where Q is the heat in calories, c is the heat capacity (1.0 for water), M is the mass (25.0 gm) of the substance, and ΔT is the difference in temperature (100 - 0 = 100 °C). So

$$Q = 1.0 \times (25.0 \text{ gm}) \times (100 \text{ °C}) = 2500 \text{ calories}$$

But something different happens when you melt or boil a substance. You keep adding heat but the temperature doesn't change. Why? Well, it takes energy (heat) to overcome (break) the inter-molecular bonds between the molecules, and the temperature cannot go up until all of these bonds are broken. So all the heat is going into breaking the inter-molecular bonds.

The amount of heat you need to melt a solid (overcome its inter-molecular bonds) is called the **heat of solidification**, written h_s. Because of **conservation of energy** (energy is not created or destroyed in a physical change), the heat you need to melt a solid is equal to the heat given off when you lower the temperature and it freezes again. Both of them are written h_s.

Likewise, the amount of heat you need to evaporate a liquid (overcome its inter-molecular bonds) is called the **heat of vaporization**, written h_v. And because of conservation of energy again, the heat you need to evaporate a liquid is equal to the heat given off when you lower the temperature and it condenses again. Both of them are written h_v.

For water, h_s is 80 calories per gram and h_v is 540 calories per gram. These values are different for other materials.

So how much heat is required to boil 25.0 gm of water? Now the equation is

$$Q = h_v \times M$$

where Q and M are the same as in the first equation and h_v is the heat of vaporization for water (540 cal/gm). So

$$Q = 540 \times (25.0 \text{ gm}) = 13,500 \text{ calories}$$

If you had tried to use the first equation, ΔT would be zero (the temperature does not change until all the inter-molecular bonds are broken) and you would get $Q = 0$. That's your clue that you need to use this equation instead.

Sometimes you need to use both equations. It's only a little harder. How much energy (heat) is given off if you cool 25.0 gm of water from steam at 120 °C to ice at -10 °C?

To answer that question, you need to break up the problem into the five parts shown in Figure 10-6 because the c's and h's are all different for each part.

Part A: 25.0 gm of steam is cooled from 120 to 100 °C   (c=0.5)

Part B: 25.0 gm of water condenses into a liquid at 100 °C   (h_v = 540)

Part C: 25.0 gm of water is cooled from 100 to 0 °C   (c=1.0)

Part D: 25.0 gm of water freezes into ice at 0 °C   (hS = 80)

Part E: 25.0 gm of ice is cooled from 0 to -10 °C   (c=0.5)

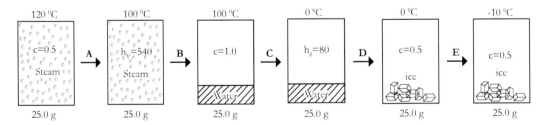

Figure 10-6

Parts A, C, and E all use the same equation $Q = c \times M \times \Delta T$, but they need to be done separately because they have different heat capacities (remember that c for a gas or solid is about half of what it is for a liquid). Likewise Parts B and D use the same equation $Q = h \times M$, but they also need to be done separately because h_v and h_s are different.

In all five parts M is 25.0 gm because of **conservation of mass** (matter is not created or destroyed in a physical change). In Part A, the heat capacity c is about 0.5, and ΔT is $120 - 100 = 20$ °C. In Part C, the heat capacity c is 1.0, and ΔT is $100 - 0 = 100$ °C. In Part E, the heat capacity c is about 0.5, and ΔT is $0 - (-10) = 10$ °C. Here you use the $Q = c \times M \times \Delta T$ equation.

What is a Gas? A Liquid? A Solid?

Part A: $Q_A = c \times M \times \Delta T = 0.5 \times (25 \text{ gm}) \times (20 \text{ °C}) = 250$ calories

Part C: $Q_C = c \times M \times \Delta T = 1.0 \times (25 \text{ gm}) \times (100 \text{ °C}) = 2{,}500$ calories

Part E: $Q_E = c \times M \times \Delta T = 0.5 \times (25 \text{ gm}) \times (10 \text{ °C}) = 125$ calories

In Part B and Part D, $\Delta T = 0$, so you need to use the other equation, and $Q = h \times M$. Here M is still 25.0 gm, but $h_V = 540$ and $h_S = 80$.

Part B: $Q_B = h_V \times M = 540 \times (25 \text{ gm}) = 13{,}500$ calories

Part D: $Q_D = h_S \times M = 80 \times (25 \text{ gm}) = 2{,}000$ calories

The total heat given off is $Q_A + Q_B + Q_C + Q_D + Q_E = 18{,}375$ calories. That wasn't so hard, was it?

Homework Problems

1. What is the gas that has a density of 1.429 gm/L at STP?

   a. N_2

   b. Ar

   c. He

   d. O_2

   e. None of the above

2. What is the gas that has a density of 1.251 gm/L at STP?

   a. CO

   b. H_2O

   c. NO

   d. CO_2

   e. None of the above

3. If you have 3.0 L of He at a pressure of 1.1 atm, what will be its volume be when the pressure changes to 0.95 atm but the temperature stays the same?

   a. 4.3 L

   b. 3.5 L

   c. 3.8 L

   d. 4.3 L

   e. None of the above

4. If you have 6.0 L of H_2 at a pressure of 0.9 atm, what will be its volume be when the pressure changes to 1.2 atm but the temperature stays the same?

   a. 3.7 L

   b. 4.5 L

   c. 3.9 L

   d. 4.3 L

   e. None of the above

5. If it is 27 °C and 1.0 atm today, what will be the volume of 5.0 L of N_2 tomorrow when it is at 24 °C and 0.9 atm?

   a. 4.7 L

   b. 5.3 L

   c. 4.9 L

   d. 5.5 L

   e. None of the above

What is a Gas? A Liquid? A Solid?

6. If it is 22 °C and 0.9 atm today, what will be the volume of 3.0 L of Ar when it is at 0 °C and 1.0 atm (STP)?

   a. 2.7 L

   b. 3.3 L

   c. 2.5 L

   d. 3.5 L

   e. None of the above

7. How much is 740 mm of mercury in atmospheres?

   a. 0.88 atm

   b. 0.99 atm

   c. 0.82 atm

   d. 0.97 atm

   e. None of the above

8. How much is 610 mm of mercury in atmospheres?

   a. 0.88 atm

   b. 0.92 atm

   c. 0.80 atm

   d. 0.77 atm

   e. None of the above

9. Which of these compounds has the strongest intermolecular forces?

   a. NO

   b. KCl

   c. HF

   d. CO_2

   e. None of the above

10. Which of these compounds has the weakest intermolecular forces?

    a. CO

    b. H_2O

    c. NaCl

    d. CCl_4

    e. None of the above

11. Which of these compounds has the lowest vapor pressure?

    a. HOH

    b. O_2

    c. CO_2

    d. H_2

    e. None of the above

12. Which of these compounds has the highest vapor pressure?

    a. HOH

    b. $HC_2H_3O_2$

    c. N_2

    d. HCl

    e. None of the above

13. Which of these compounds has the lowest surface tension?

    a. HCl

    b. C_6H_6

    c. HOH

    d. $HC_2H_3O_2$

    e. None of the above

14. Which of these compounds has the highest surface tension?

    a. $HC_2H_3O_2$

    b. C_6H_6

    c. HOH

    d. CCl_4

    e. None of the above

15. Which of these compounds has the highest viscosity?

    a. $C_6H_{12}O_6$

    b. Hg_2

    c. CCl_4

    d. Br_2

    e. None of the above

What is a Gas? A Liquid? A Solid?

16. Which of these compounds has the lowest viscosity?

    a. CCl_4

    b. HOH

    c. $C_6H_{12}O_6$

    d. $HC_2H_3O_2$

    e. None of the above

17. Which of these compounds has the lowest melting point?

    a. HOH

    b. $HC_2H_3O_2$

    c. N_2

    d. HCl

    e. None of the above

18. Which of these compounds has the highest melting point?

    a. HOH

    b. O_2

    c. CO_2

    d. H_2

    e. None of the above

19. What kind of crystalline solid is NaCl?

    a. Molecular

    b. Covalent

    c. Ionic

    d. Metallic

    e. None of the above

20. What kind of crystalline solid is Ag?

    a. Molecular

    b. Covalent

    c. Ionic

    d. Metallic

    e. None of the above

21. What is the heat capacity of a substance if it takes 0.50 calorie to heat 0.25 gm of it by 1 °C?

   a. 0.25

   b. 1.00

   c. 0.50

   d. 2.00

   e. None of the above

22. What is the heat capacity of a substance if it takes 0.25 calorie to heat 0.50 gm of it by 1 °C?

   a. 0.25

   b. 1.00

   c. 0.50

   d. 2.00

   e. None of the above

23. How much heat is needed to vaporize 50 gm of liquid ethanol (c = 0.59) at 10 °C if its boiling point is 78 °C and its heat of vaporization (h_v) is 200 cal/gm?

   a. 11,000 cal

   b. 11,700 cal

   c. 13,400 cal

   d. 12,000 cal

   e. None of the above

24. How much heat is needed to heat 20 gm of ice (c = 0.5) at -25 °C to superheated steam (c = 0.5) at 125 °C?

   a. 10,500 cal

   b. 14,900 cal

   c. 15,200 cal

   d. 13,700 cal

   e. None of the above

Chapter 11

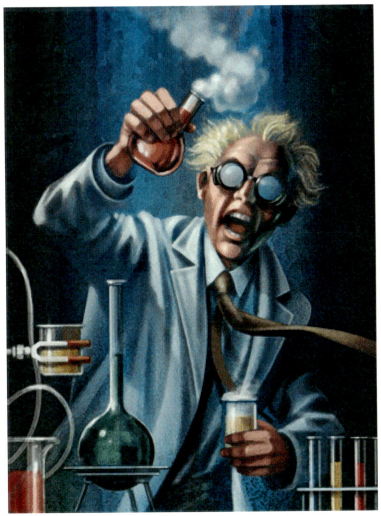

Credit -http://www.ttfatloss.com/wp-content/uploads/2009/03/mad_scientist.gi

What are Solutions?

You might think you already know what a solution is, but actually the subject is more complicated than you think. It's not just mixing rum and coke to make a drink. As you will see, it's way more than that.

Solutes and Solvents

If you combine salt and water, which one is the solvent? The water? It actually depends on how much salt you have. If you have more salt than water, then the salt is the solvent. The substance with the larger mass is always the **solvent**, even if it's not a liquid. The substance with the smaller mass is always the **solute**, even if it's not a solid.

We usually think of **solutions** as two liquids mixed together, but actually they come in all forms. You can have a gas mixed with a gas, a gas mixed with a liquid, a gas mixed with a solid, a liquid mixed with a liquid, a liquid mixed with a solid, and even a solid mixed with a solid. Solutions are usually homogeneous (the same everywhere you look).

Table 11-1: Different Types of Solutions

SOLUTE	SOLVENT	EXAMPLE	COMPOSITION
Gas in a	Gas	Pure air	Oxygen in Nitrogen
	Liquid	Soda pop	CO_2 in Sugar Water
	Solid	Saw dust	Air in Wood
Liquid in a	Gas	Moist air	Water in Air
	Liquid	Martini	Water in Gin
	Solid	Potting soil	Water in Dirt
Solid in a	Gas	Dust storm	Dirt in Air
	Liquid	Sea water	Salt in Water
	Solid	Sterling silver	Copper in Silver

Mass Percent

You sometimes hear some people use very vague terms when they talk about the **concentration** (strength) of solutions like acids or bases. You hear them say that the acid or base is "weak" or "strong". There are much better ways to describe the concentration of a solution. One of those is mass percent.

Mass percent is the mass of the solute divided by the mass of the solution and written as a percent.

$$\text{Mass Percent} = \frac{\text{Mass of solute}}{\text{Mass of solution}} \times 100\%$$

But be careful! The **mass of the solution** is not just the mass of the solvent. It is the mass of the solute added to the mass of the solvent.

(Grams of solution) = (Grams of solute) + (Grams of solvent)

What are Solutions?

Try an example. What is the mass percent of 6.5 gm of salt dissolved in 93.5 gm of water?

$$\text{Mass Percent} = \frac{6.5 \text{ gm}}{6.5 \text{ gm} + 93.5 \text{ gm}} \times 100\% = 6.5\%$$

Molarity

The concentration of a solution can also be expressed by molarity. **Molarity** is the number of moles of solute per liter of solution.

$$\text{Molarity} = \frac{\text{(Moles of solute)}}{\text{(Liters of solution)}}$$

Here you need to use the MM, MV, or MN from Chapter 8 to find the moles of the solute. The liters of solution is just the volume of the solution in liters.

Be careful again! The volume of the solution is usually not the volume of the solute plus the volume of the solvent. If the solute dissolves completely, the volume of the solution stays the same as the volume of the solvent.

Try an example. What is the molarity of a solution prepared by completely dissolving exactly 100 gm of NaOH in exactly 250 mL (0.250 L) of water?

Since the NaOH dissolves completely, the volume of the solution stays 0.250 L. Using its MM (40.00 gm per mole), the moles of NaOH is

$$\frac{100 \text{ gm NaOH}}{40.00 \text{ gm per mole}} = 2.50 \text{ moles NaOH}$$

$$\text{Then} \quad \text{Molarity} = \frac{2.50 \text{ moles NaOH}}{0.250 \text{ L of solution}} = 10.0\text{M NaOH}$$

By the way, a solution made like this where you carefully weigh a mass of solute and completely dissolve it in a carefully measured volume of water is called a standard solution. A **standard solution** is one for which the concentration is very accurately known (to at least 3 significant digits). Standard solutions are used for doing very accurate chemical analysis.

You can also use the equation for molarity in reverse. Suppose you want to make 500 mL of a 12M solution of NaOH. How much NaOH do you need?

Solving the molarity equation for moles

$$\text{(Moles of solute)} = \text{(Molarity)} \times \text{(Liters of solution)} = 6.0 \text{ moles NaOH}$$

Now, using the MM for NaOH (40.0 gm per mole), you get

$$6.0 \text{ moles} \times 40.0 \text{ gm per mole} = 240 \text{ gm NaOH}$$

You need to weigh out 240 gm of NaOH and dissolve it in 500 mL of DI (de-ionized) water. Always use DI water to make aqueous solutions!

Solubility

Solubility is how much of one substance can be mixed with another to form a solution. Some substances are more soluble than others. NaCl (salt), for example, is very soluble in water, but Cu (copper) is not. Solubility is usually given as grams of solute in 100 gm of solvent, but if the solvent is a gas or a liquid, it can also be given as grams of solute in 100 mL of solvent.

Solubility and Temperature

The solubility of a substance can increase or decrease due to several things. One such thing is temperature. The solubility of a gas in a liquid decreases as the temperature increases. You see that when you open a hot can of soda. It fizzes over! The same can of soda only bubbles when you open it cold. So the CO_2 in the sugar water is less soluble when hot than when it is cold.

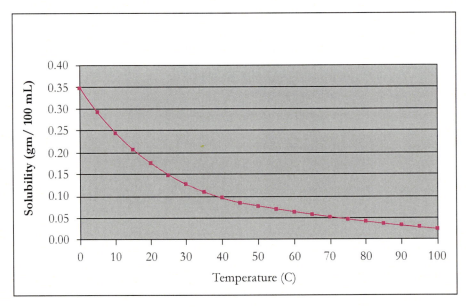

Figure 11-2: Solubility of Carbon Dioxide in Water

Likewise, the solubility of a gas in a solid also decreases as the temperature increases. The reason for that is the same as for a gas in a liquid. The gas expands as it gets hotter, and some of it is pushed out of the solid or liquid.

On the other hand, the solubility of a solid in a liquid increases as the temperature increases. You see that when you need to add twice as many chlorine tablets to your swimming pool in the summer than in winter. So the chlorinating tablets are more soluble in hot water than in cold water. You will see the reason for that later on when we talk about the rate of dissolving.

What are Solutions?

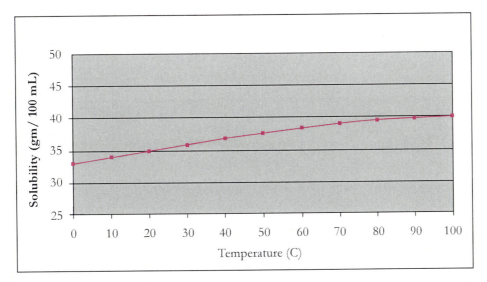

Figure 11-3: Solubility of NaCl in Water

The solubility of a gas in a gas, a liquid in a liquid, or a solid in a solid does not change very much as the temperature increases because the solute expands at nearly the same rate as the solvent.

Solubility and Pressure

Pressure (push per square inch) is something else that can increase the solubility of a substance. A solid or liquid, as you already know, cannot be compressed. So the solubility does not change with an increase in pressure.

A gas, on the other hand, can be compressed a lot. For a gas in a liquid or a gas in a solid, the solubility increases as the pressure increases because the gas is literally pushed back into the liquid or solid.

This equation shows the change in solubility when you change the pressure.

$$S_2 = \frac{P_2}{P_1} \times S_1$$

where P_1 and S_1 are the original pressure and solubility and P_2 and S_2 are the new pressure and solubility. This equation says that if you double the pressure, you double the solubility.

Try an example. If the solubility of CO_2 in water is 0.35 gm per 100 mL at STP, what will it be at the same temperature and 3 atmospheres pressure? Remember that STP means $T_1 = 0°$ C and P_1 = 1 atmosphere. You are told that S_1 = 0.35 and P_2 = 3 atm, so using the equation you just saw

$$S_2 = \frac{3 \text{ atm}}{1 \text{ atm}} \times 0.35 = 1.05 \text{ gm per } 100 \text{ mL}$$

In this case the temperature stayed the same, so you didn't need to worry about what difference the temperature would make. But if the temperature had not stayed the same, you could use the ideal gas law to estimate the new pressure (P_2) at the new temperature. Remember how to do that?

Unsaturated and Supersaturated Solutions

So far we have been talking about **saturated** solutions. They are solutions where the solute dissolves as much as it can (to its maximum possible concentration). But if you add too little solute, you can make what is called an **unsaturated** solution. Under certain conditions you can even make a **supersaturated** solution where the concentration of the solute is greater than what would normally be possible. Supersaturated solutions are unstable and precipitate out part of the solute at the drop of a pin.

As you already saw, the solubility of a substance changes with temperature. The curve in Figure 11-3 is for a saturated solution. The area below the curve is unsaturated and has less solute than it can hold, and the area above the curve is supersaturated and has more solute than it can normally hold. The **saturation curve** in Figure 11-4 shows that.

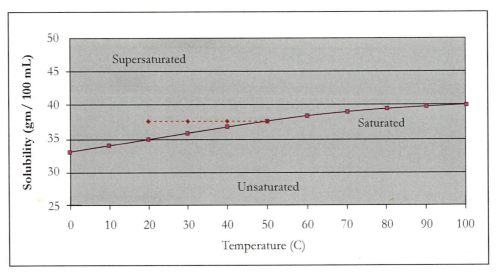

Figure 11-4: Seturation Curve for NaCl in Water

If the solution is cooled very slowly and no one moves or makes any noise, the concentration of the solute will stay the same as the temperature drops, and you make a supersaturated solution. Look at the dotted line in Figure 11-4. It shows what happens if the solution is cooled from 50 to 20 °C very slowly. But if someone makes a noise or drops something, enough NaCl will precipitate out to bring the concentration of the solution back down to the saturated curve where it should be.

Process of Dissolving

Like dissolves like. Polar solvents will dissolve polar solutes, but they will not dissolve non-polar solutes. Likewise, non-polar solvents will dissolve non-polar solutes, but they will not dissolve polar solutes. Why is that?

Non-polar covalent compounds, if you remember, are compounds where the bonding electrons are shared equally or nearly equally. So there is no dipole because there is no difference in charge between the atoms of the compound. Polar covalent compounds, on the other hand, are compounds where the bonding electrons are not shared equally. That results in a δ^+ charge at one end of the molecule and a δ^- charge at the other. For purposes of dissolving, ionic compounds are thought of as polar because they have a +1 charge at one end and a −1 charge at the other.

When a polar solute is put into a polar solvent, the δ^+ or δ^- charges of one pull on the δ^- or δ^+ charges of the other. In the case of ionic compounds like NaCl in a polar solvent like H_2O, the Na^+ and Cl^- ions are actually pulled apart (**dissociated**) by the water molecules. The Cl^- ions are attracted to the H^+ sides of some of the water molecules, and the Na^+ ions are attracted to the O^{-2} sides of others. The Na^+ and Cl^- ions each get surrounded by a sort of cage of water molecules. That cage is called a **solvent cage**.

With polar covalent molecules dissolved in polar solvents, the result is not as extreme, but the molecules still arrange themselves (**ionize**) in such a way that the positive sides of the solute face the negative sides of the solvent and vice versa. Here it is more like a **solvent blanket** than a solvent cage.

What happens when you add non-polar molecules to a polar solution? Well the polar molecules are still attracted to each other and will slide past the non-polar molecules until all the polar molecules are in one place and all the non-polar molecules are left behind.

If the polar molecules are less dense than the non-polar molecules, they will float on top of the non-polar molecules. If the polar molecules are more dense than the non-polar molecules, it will be the other way around. Have you ever seen vinegar and oil separate into two layers? That's why.

When two liquids separate like that, they are called **immiscible** (do not mix). If they do mix (when like dissolves like), they are called **miscible**.

How can you tell which solvents are polar? As a rule of thumb, if the solvent contains oxygen, it is polar. H_2O is polar and so is $HC_2H_3O_2$ (vinegar), C_2H_5OH (ethanol), CH_3OH (methanol), and C_3H_6O (acetone). Ether, $C_4H_{10}O$, is an exception. Despite having oxygen, it is non-polar.

Rate of Dissolving

It can take up to several minutes for a substance to dissolve completely, but there are a few things you can do to speed it up.

If it is a solid, you can grind up the solute. That will increase the exposed surface of the solute and let the solvent molecules attack more of the solute at the same time. That will speed things up.

You can stir the solution. That will move the saturated solvent away from the solute and bring more of the unsaturated solvent near the solute. That will speed things up.

You can also increase the temperature. That will increase the energy of the solution, and that will increase the rate at which solvent molecules collide with solute molecules. That will speed things up. By the way, this also explains why the solubility of solids in liquids increases with temperature.

Dilution of Solutions

Acids are commonly sold in 5 liter glass jars at concentrations up to 18M. At that strength, H_2SO_4 (sulfuric acid) will eat through a leather shoe in less than 10 seconds! It is dangerous to use concentrated acids like that in a laboratory, so they are usually diluted down to a concentration of about 1M.

So how do you dilute a solution? Well, you know that the amount of the solute will not change by just adding more water, so

$$\text{Moles}_2 \text{ (of solute)} = \text{Moles}_1 \text{ (of solute)}$$

where the subscripts 1 and 2 refer to before and after the dilution. Now use the equation for molarity and solve it for moles of solute.

$$(\text{Moles of solute}) = (\text{Molarity}) \times (\text{Liters of solution})$$

$$\text{So } M_2 \times V_2 = M_1 \times V_1$$

where M_1 and V_1 are the original molarity and volume and M_2 and V_2 are the final molarity and volume.

Try an example. You have a jug of 18M H_2SO_4 but need to make 450 mL of 2M sulfuric acid. How much of the 18M acid should you use? Solving that equation for V_1, you get

$$V_1 = \frac{M_2}{M_1} \times V_2 = \frac{2}{18} \times 450 \text{ mL} = 50 \text{ mL}$$

But be careful! When using concentrated acids, always add the acid to the water, not the water to the acid! The same goes for concentrated bases.

So add 50 mL of the 18M sulfuric acid to 400 mL of DI water. That will make 450 mL of 2M sulfuric acid.

See all you can do now? You're getting close to being a chemist!

What are Solutions?

Homework Problems

1. Which of the following is not a solution?

    a. Brass

    b. Coke

    c. Tin

    d. Dental amalgam

    e. None of the above

2. Which of the following is not a solution?

    a. Charcoal

    b. Salt water

    c. Vinegar

    d. Sterling silver

    e. None of the above

3. Which is the solvent if you have 71 gm of salt in 69 gm of water?

    a. Sa

    b. Na

    c. NaCl

    d. HOH

    e. None of the above

4. Which is the solute if you have 62 gm of salt in 60 gm of water?

    a. NaCl

    b. Na

    c. HOH

    d. Sa

    e. None of the above

5. What is the mass percent of the solution if you have 44 gm of sodium dissolved in 65 gm of water?

    a. 68 %

    b. 33 %

    c. 62 %

    d. 40 %

    e. None of the above

6. What is the mass percent of the solution if you have 9 gm of sand dissolved in 9 gm of water?

   a. 30 %

   b. 72 %

   c. 50 %

   d. 43 %

   e. None of the above

7. What is the mass percent of the solution if you have 3.9 gm of potassium chloride dissolved in 3.9 gm of pure water?

   a. 99 %

   b. 75 %

   c. 50 %

   d. 25 %

   e. None of the above

8. What is the mass percent of the solution if you have 15 gm of sodium nitrate dissolved in 85 gm of de-ionized water?

   a. 10 %

   b. 15 %

   c. 18 %

   d. 23 %

   e. None of the above

9. What is the mass percent of the solution if you have 19 gm of oxygen in 95 gm of air?

   a. 14 %

   b. 17 %

   c. 20 %

   d. 23 %

   e. None of the above

10. What is the mass percent of the solution if you have 5 gm of sodium acetate in 15 gm of de-ionized water?

    a. 25 %

    b. 33 %

    c. 65 %

    d. 30 %

    e. None of the above

What are Solutions?   133

11. What is the molarity of the solution if you have 5.0 gm of calcium dissolved in 250 mL of pool water?

   a. 0.8 M

   b. 2.5 M

   c. 1.2 M

   d. 0.5 M

   e. None of the above

12. What is the molarity of the solution if you have 8.0 gm of strontium dissolved in 100 mL of liquid ammonia?

   a. 0.5 M

   b. 1.0 M

   c. 2.5 M

   d. 1.4 M

   e. None of the above

13. What is the molarity of the solution if you have 13.7 gm of barium dissolved in 100 mL of liquid ammonia?

   a. 1.0 M

   b. 1.5 M

   c. 2.0 M

   d. 2.5 M

   e. None of the above

14. What is the molarity of the solution if you have 8.0 gm of oxygen dissolved in 125 mL of liquid nitrogen?

   a. 3.5 M

   b. 2.8 M

   c. 2.0 M

   d. 1.5 M

   e. None of the above

15. What is the molarity of the solution if you have 11.5 gm of sodium acetate dissolved in 250 mL of water?

   a. 5.0 M

   b. 3.5 M

   c. 2.5 M

   d. 2.0 M

   e. None of the above

16. What is the molarity of the solution if you have 7.0 gm of nitrogen dissolved in 500 mL of distilled water?

    a. 0.5 M

    b. 1.0 M

    c. 2.0 M

    d. 2.5 M

    e. None of the above

17. If a glass of soda can hold 1.76 gm of CO_2 per liter at 20 °C, how much can it hold at 22 °C?

    a. about 10% more

    b. about two times more

    c. about 10% less

    d. about two times less

    e. None of the above

18. If a swimming pool can hold 0.182 gm of Cl_2 per liter at 22 °C, how much can it hold at 20 °C?

    a. about 10% more

    b. about two times more

    c. about 10% less

    d. about two times less

    e. None of the above

19. If a swimming pool can hold 0.182 gm of Cl_2 per liter at one atmosphere of pressure, how much can it hold at two atmospheres?

    a. about 10% more

    b. about two times more

    c. about 10% less

    d. about two times less

    e. None of the above

20. If a glass of soda can hold 1.76 gm of CO_2 at one atmosphere of pressure, how much can it hold at two atmospheres?

    a. about 10% more

    b. about two times more

    c. about 10% less

    d. about two times less

    e. None of the above

What are Solutions?

21. To how much distilled water should you add 50 mL of 12 molar HCl to get a 2 M solution?

    a. 150 mL

    b. 200 mL

    c. 250 mL

    d. 300 mL

    e. None of the above

22. To how much distilled water should you add 100 mL of 10 molar HNO_3 to get a 1 M solution?

    a. 750 mL

    b. 900 mL

    c. 950 mL

    d. 1000 mL

    e. None of the above

23. How much 12 M NaOH do you need to add to 500 mL of de-ionized water to make a 2 M solution of NaOH?

    a. 500 mL

    b. 100 mL

    c. 450 mL

    d. 300 mL

    e. None of the above

24. How much 15 M NH_4OH do you need to add to 700 mL of pure water to make a 1 M solution of NH_4OH?

    a. 50 mL

    b. 35 mL

    c. 70 mL

    d. 25 mL

    e. None of the above

Chapter 12

Credit: http://sciencegames.4you4free.com/acids_bases_ph_scale.gif

What is an Acid? A Base?

When we talked about acids and bases before, we said that an acid is a compound that starts with an H (hydrogen) and a base is a compound that ends with an OH (hydroxide). But actually the subject of what is an acid and what is a base is a lot more complicated. The H and OH definitions are only one way of looking at acids and bases. There are others.

Arrhenius Acids and Bases

Those definitions of acids and bases are from a chemist named Arrhenius. An **Arrhenius acid** is a polar compound that ionizes (partially separates) in water to give off H^+ ions. Strong acids ionize a lot and weak acids only a little. HCl ionizes almost 100%, so it makes a lot of H^+ ions and is a strong acid. $HC_2H_3O_2$ ionizes less than 10%, so it makes a lot less H^+ ions and is a relatively weak acid. Other strong and weak acids are listed in Table 12-1.

Table 12-1: Some Strong and Weak Acids

ACID	FORMULA	IONIZATION	STRENGTH
Hydrocloric acid	HCl	~100%	Strong
Nitric acid	HNO_3	~100%	Strong
Sulfuric acid	H_2SO_4	~100%	Strong
Carbonic acid	H_2CO_3	<10%	Weak
Phosphoric acid	H_3PO_4	<10%	Weak
Acetic acid	$HC_2H_3O_2$	<10%	Weak

Likewise, an **Arrhenius base** is an ionic compound that dissociates (separates completely) in water to give off OH^- ions. Strong bases dissociate a lot and weak bases only a little. NaOH dissociates almost 100%, so it makes a lot of OH^- ions and is a strong base. NH_4OH dissociates less than 10%, so it makes a lot less OH^- ions and is a relatively weak base. Other strong and weak bases are listed in Table 12-2.

Table 12-2: Some Strong and Weak Bases

BASE	FORMULA	IONIZATION	STRENGTH
Sodium hydroxide	NaOH	~100%	Strong
Potassium hydroxide	KOH	~100%	Strong
Calcium hydroxide	$Ca(OH)_2$	~100%	Strong
Zinc hydroxide	$Za(OH)_2$	<10%	Weak
Magnesium hydroxide	$Mg(OH)_2$	<10%	Weak
Ammonium hydroxide	NH_4OH	<10%	Weak

By the way, the Arrhenius definitions say that HOH (water) is both an acid and a base because it both starts with an H and ends with an OH. That makes water unique and is why water has such a special place in chemistry.

You may have noticed that acids (polar compounds) **ionize**, but bases (ionic compounds) **dissociate**. The ions in both of them separate to some extent, but as you saw when discussing solutions, they separate in different ways. Chemists sometimes make these small distinctions. Humor them.

Bronsted and Lowry Acids and Bases

The Bronsted and Lowry (B-L) definitions of acids and bases are different. They apply to more solvents than just water, and they see acids and bases in compounds that you wouldn't think of as acids and bases. A **B-L acid** is any substance that donates H^+ (proton) ions to another substance. So it is a **proton donor**. A **B-L base** is any substance that accepts H^+ ions. So it is a **proton acceptor**. That's the same thing Arrhenius said. Right? Well you need to look at some examples to see the difference.

Neutralization Reactions

Remember that a neutralization reaction is a special type of double replacement reaction where the reactants are an acid and a base and the products are a salt and water. First look at this neutralization reaction from the Arrhenius point of view.

$$HCl\ (aq)\ +\ NaOH\ (aq)\ \rightarrow\ NaCl\ (aq)\ +\ HOH\ (l)$$

The HCl ionizes to give off H^+ ions. So it is an Arrhenius acid. The NaOH dissociates to give off OH^- ions. So it is an Arrhenius base.

Now look at another reaction from the Arrhenius point of view.

$$HCl\ (aq)\ +\ NH_3\ (aq)\ \rightarrow\ NH_4Cl\ (aq)$$

The HCl is an Arrhenius acid because it gives off H^+ ions, but the NH_3 is not an Arrhenius base because it does not give off OH^- ions. According to Arrhenius, this reaction is not a neutralization reaction.

But now let's look at the same two reactions from the B-L point of view. In the first reaction the HCl donates an H^+ ion, so it is a B-L acid. The NaOH accepts the H^+ ion, so it is a B-L base. No difference here.

Now look at the second reaction. Again the HCl donates an H^+ ion, so it is a B-L acid. But the NH_3 accepts an H^+ ion to make NH_4^+, so NH_3 is a B-L base. Arrhenius says NH_3 is not a base, but Bronsted and Lowry say it is. So who are you going to believe?

Most chemists use the B-L definitions, but the Arrhenius definitions are helpful when you are just starting to learn chemistry!

The B-L definitions lead to an interesting result. Look at these reactions:

$$NaHCO_3\ (aq)\ +\ NaOH\ (aq)\ \rightarrow\ Na_2CO_3\ (aq)\ +\ HOH\ (l)$$
$$HCl\ (aq)\ +\ NaHCO_3\ (aq)\ \rightarrow\ NaCl\ (aq)\ +\ H_2CO_3\ (aq)$$

In the first reaction, $NaHCO_3$ donates an H^+ ion to NaOH to make HOH. So it is a B-L acid. In the second reaction, $NaHCO_3$ accepts an H^+ ion from HCl to make H_2CO^3. So it is a B-L base. First $NaHCO_3$ is an acid, and then it is a base! Substances that can both donate and accept protons like that are called **amphiprotic**. Most compounds with HCO_3^- or HSO_4^- ions are amphiprotic, but there are other compounds that are also amphiprotic.

Measuring the Strength of an Acid or Base

Acidity is measured using the pH scale. Anyone who has a swimming pool knows about measuring the pH of the water in their pool. But what they probably do not know is that pH is measured on a logarithmic scale where each number is 10 times stronger than the previous one. It's like the Richter scale that measures earthquakes. Mathematically it is written like this,

$$pH = -\log [H^+]$$

where $[H^+]$ is the hydrogen ion concentration (molarity). The minus sign turns the scale upside down so that 1 becomes stronger than 2 and so on.

Remember strong acids ionize nearly 100%, but weak acids ionize less than 10%. So strong acids give off many more H^+ ions than weak acids do. If you use some real numbers in the pH equation, you will find that the pH of a strong acid is around 1 while the pH of a weak acid is more like 4.

When you get into bases, the OH^- ions use up most of the H^+ ions to make water (HOH). So the H^+ ion concentration gets much smaller. A weak base now has a pH of around 10 while a strong base has a pH more like 13.

So acids and bases have a pH between 0 and 14. Water, which is both, is said to be neutral and has a pH of 7, right in the middle. Table 12-3 shows the pH of some acids and bases calculated using the pH equation.

Table 12-3: The pH of Acids and Bases

Strength	pH	Ionization	Examples
Strong base	14 13 12	~100%	Drain cleaner
Weak base	11 10 9	<10%	Soap
Nautral	8 7 6	<1%	Pure water
Weak acid	5 4 3	<10%	Vinegar
Strong acid	2 1 0	~100%	Stomach acid

pH Indicators

In the laboratory you don't actually measure the H+ ion concentration of a solution to tell whether it is an acid or base. Instead you use **pH indicators**, which are substances that change color when put into an acid or base.

There are four pH indicators commonly used. **Methyl red** is (you guessed it) red at a pH below 5, and starts turning yellow at 5. Another indicator is **bromothymol blue**. It is yellow below 7 and starts turning blue at 7.

The most commonly used pH indicator is **phenolphthalein**. It is clear at a pH below 9 and starts turning red at 9. Phenolphthalein is the liquid you add to a sample of pool water when you measure its pH.

Why use phenolphthalein which changes color at a pH of 9? The line separating acids and bases is at 7. Why not use bromothymol blue instead? One answer is that the difference between a pH of 7 and 9 is only a drop or two of a base solution. Another answer is cost. Phenolphthalein is cheaper.

A fourth indicator often used in the laboratory is **litmus paper**. It doesn't really measure the pH, but only tells whether a solution is acidic (an acid) or basic (a base). Blue litmus paper turns red when dipped into an acidic solution, and red litmus paper turns blue when dipped into a basic solution.

Titration of an Acid with a Base

You can determine the molarity of an acid solution using a pH indicator and a technique called titration. **Titration** is done using a special graduated cylinder with a valve at the bottom. It is called a **burette**.

You start with a base solution which you make up yourself, so you know its molarity. You pour some of this base solution into the burette and record its starting volume using the lines marked on the side of the burette. Later on you read the final volume, and the difference is how much of the base you used.

Underneath the valve at the bottom of the burette you put a beaker with an acid of unknown molarity. Before pouring the acid into it, you measure the volume using a graduated cylinder. So you know the volume of the acid.

Now you add a drop of phenolphthalein to the beaker with the acid. Then you add drops of the base solution to the acid until it turns red. That point is called the endpoint. At **endpoint**, the moles of acid equal the moles of base.

The procedure is better explained by doing an example. Say you add 40.0 gm of NaOH to 1.0 L of DI water to make a 1.0 M NaOH solution, and then you pour 90 mL of it into a burette. Next you pour 50 mL (0.050 L) of vinegar ($HC_2H_3O_2$) into the beaker underneath. You add a drop of phenolphthalein to the beaker, and then start adding NaOH drop by drop using the valve at the bottom of the burette. After several minutes of doing this, the solution turns red, and you read the amount of NaOH left over. It is

65 mL, so you used 90 – 65 = 25 mL (0.025L) of the NaOH base to neutralize the 50 mL of the $HC_2H_3O_2$ acid. This is the reaction that happened,

$$HC_2H_3O_2 \text{ (aq)} + NaOH \text{ (aq)} \rightarrow NaC_2H_3O_2 \text{ (aq)} + HOH \text{ (l)}$$

From the known molarity and volume of the NaOH, you can use the definition for molarity to find the number of moles of the base solution,

$$\text{Moles} = \text{Molarity x Liters} = (1M) \times (0.025 \text{ L}) = 0.025 \text{ moles NaOH}$$

Since you completely neutralized the acid (at endpoint), the moles of acid equals the moles of base. Because you have 0.025 moles of NaOH, you must have 0.025 moles of $HC_2H_3O_2$. Using the definition for molarity again,

$$\text{Molarity} = \frac{\text{Moles}}{\text{Liter}} = \frac{0.025 \text{ mol}}{0.050 \text{ L}} = 0.50M \ HC_2H_3O_2$$

So the previously molarity of the acid is 0.50M. Not so hard, is it?

If you use an acid of known molarity, you can use the same technique to find the molarity of a base.

Determining the pH of an Acid or Base

Using the pH equation, you can calculate the pH for any solution if you know the concentration of the H^+ ions it contains. You can use the titration method you just learned to find the H^+ ion concentration (the molarity).

Say you measure the H^+ ion concentration of a cup of coffee using titration and find it is 0.00001M. What is its pH? Using the log key on a calculator,

$$pH = -\log [H^+] = -\log (.00001) = -(-5) = 5$$

Try another example. Using titration you find the H^+ ion concentration of a cup of liquid bleach is 1.0×10^{-12}. What is its pH? Using the log key again,

$$pH = -\log (1.0 \times 10^{-12}) = -(-12) = 12$$

So the coffee is a weak acid, but the bleach is a fairly strong base.

Now what if you know the pH but want to know the H^+ ion concentration? If you know something about logarithms, you can invert (reverse) the pH equation into this one,

$$[H^+] = 10^{-pH}$$

So if the pH of an antacid tablet is 9, what is its H^+ ion concentration? Using the inverted pH equation and the exponent key on your calculator,

$$[H^+] = 10^{-9} = 0.000000001M$$

In some cases, the pH of a solution must be kept in a very narrow range. Solutions that can do this are called **buffers**. Human blood, for example, must be kept at a pH between

What is an Acid? A Base?

7.35 and 7.45. The buffer that does this is a weak solution of H_2CO_3 and $NaHCO_3$." The acid neutralizes an increase in basicity, while the bicarbonate cancels out an increase in acidity.

Ionization of Water

You now know that water has a pH of 7. Using the inverted pH equation,

$$[H^+] = 10^{-pH} = 10^{-7} = 0.0000001 M$$

If you think about it, that means that water (very slightly) ionizes like this,

$$HOH\ (l) \rightarrow H^+\ (aq) + OH^-\ (aq)$$

So water must have an equal number of H^+ and OH^- ions. That means,

$$[OH^-] = [H^+] = 10^{-7} M$$

When you multiply the H^+ and OH^- ion concentrations together, you get what is called the ionization constant (IC) for water,

$$K_w = [H^+] \times [OH^-] = 10^{-7} \times 10^{-7} = 10^{-14}$$

It turns out that in any aqueous solution, the product of $[H^+]$ and $[OH^-]$ is always the same and is equal to K_w. The reason for that is Nice-to-Know, so we skip it.

This IC for water leads to some interesting results. You know that in a strong acid, the concentration of H^+ ions is high. So according to the IC equation for water, the concentration of OH^- ions has to be low. What happens is that the H^+ ions swallow up most of the OH^- ions and turn them into HOH. You can calculate the new OH^- ion concentration using K_w,

$$[OH^-] = \frac{K_w}{[H^+]}$$

Take an example. Lemon juice has a pH of 2, so using the inverted pH equation, its H^+ ion concentration is 10^{-2}. Then the OH- ion concentration is,

$$[OH^-] = \frac{K_w}{[H^+]} = \frac{10^{-14}}{10^{-2}} = 10^{-12} M$$

Likewise, in a strong base solution, the concentration of OH^- ions is high, so the concentration of H^+ ions has to be low. Here the OH^- ions swallow up most of the H^+ ions and turn them into HOH. The new H^+ ion concentration is,

$$[H^+] = \frac{K_w}{[OH^-]}$$

Another example. Ammonia cleaner has a pH of 11. So $[OH^-]$ is 10^{-11} M. Then,

$$[H^+] = \frac{K_w}{[OH^-]} = \frac{10^{-14}}{10^{-11}} = 10^{-3} = 0.001 M$$

Electrolyes

When we discussed solutions you learned that if you put an ionic compound into DI water, the positive and negative ions actually dissociate (separate). Polar compounds also ionize into positive and negative ions in a solution.

When you put a voltage across an ionic solution, the negative ions in the solution act just like the negative electrons in a wire and conduct (carry) electricity through the solution. It turns out that the positive ions in the solution can also conduct electricity. So you get twice the current.

If a compound does not ionize (like non-polar compounds), of course, it cannot conduct electricity. Substances that dissociate or ionize such that they can conduct electricity are called **electrolytes**. Substances that cannot conduct electricity are called **non-electrolytes**.

If you have a strong acid or base, it ionizes nearly 100%. So it releases more ions into the solution and can carry more current. If it is a weak acid or base, it releases much less ions and can carry much less current. Solutions that can carry a lot of current are called **strong electrolytes**. Solutions that can carry only a little current are called **weak electrolytes**.

Figure 12-4 shows what happens when you put a battery and light bulb across solutions of weak electrolytes and strong electrolytes.

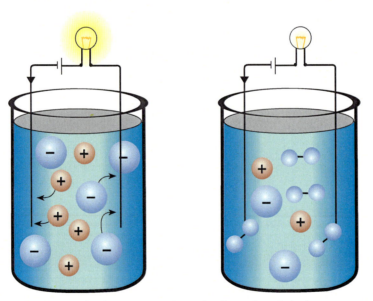

Figure 12-4: Strong and Weak Electrolyte Solutions

Ionic Equations

You can see what is happening with electrolytes in an aqueous solution by writing all the electrolytes in ionic form. The positive and negative ions are written as if they were separate compounds. Solids and liquids (non-electrolyte compounds) do not dissolve, so they are left as compounds. When you do that, you get what is called the **total ionic equation**.

What is an Acid? A Base?

Take the reaction between HCl and NaOH as an example,

$$HCl\ (aq) + NaOH\ (aq) \rightarrow NaCl\ (aq) + HOH\ (l)$$

HCl is a strong acid, so it ionizes; NaOH is a strong base, so it ionizes; and NaCl is an ionic compound, so it ionizes. But water is a very weak acid and base, so it does not really ionize. So the total ionic equation is,

$$H^+ + Cl^- + Na^+ + OH^- \rightarrow Na^+ + Cl^- + HOH$$

Can you see that Na^+ and Cl^- appear on both sides of the reaction? Ions like these are called **spectators**. Like spectators at a car accident, they make it hard to see what happened. So if you remove the spectators, you get,

$$H^+ + OH^- \rightarrow HOH$$

Now you can see what actually happened: the H^+ ions and OH^- ions combined to make HOH (water). When you remove the spectators, the equation you get is called the **net ionic equation**. Net ionic equations are useful because you can quickly see what actually happened.

You are almost a chemist. Just two more chapters!

Homework Problems

1. What is the B-L base in the following reaction?
   KCN (aq) + HF (aq) → KF (aq) + HCN (aq)
   a. KCN
   b. HF
   c. KF
   d. HCN
   e. None of the above

2. What is the B-L acid in the following reaction?
   KOH (aq) + KHCO$_3$ (aq) → K$_2$CO$_3$ (aq) + HOH (l)
   a. HOH
   b. K$_2$CO$_3$
   c. KHCO$_3$
   d. KOH
   e. None of the above

3. What is the B-L acid in the following reaction?
   NH$_3$ (aq) + HC$_2$H$_3$O$_2$ (aq) → NH$_4$C$_2$H$_3$O$_2$ (aq)
   a. C$_2$H$_3$O$_2$
   b. NH$_4$
   c. HC$_2$H$_3$O$_2$
   d. NH$_3$
   e. None of the above

4. What is the B-L base in the following reaction?
   KC$_2$H$_3$O$_2$ (aq) + HNO$_3$ (aq) → HC$_2$H$_3$O$_2$ (aq) + KNO$_3$ (aq)
   a. HNO$_3$
   b. KNO$_3$
   c. KC$_2$H$_3$O$_2$
   d. HC$_2$H$_3$O$_2$
   e. None of the above

5. Is it an acid or a base if the pH is 8?
   a. neither
   b. base
   c. both
   d. acid
   e. None of the above

What is an Acid? A Base?

6. Is it an acid or a base if the pH is 7?

   a. base

   b. both

   c. acid

   d. neither

   e. None of the above

7. Is it an acid or a base if the pH is 6?

   a. base

   b. both

   c. acid

   d. neither

   e. None of the above

8. Is it an acid or a base if the pH is 8?

   a. neither

   b. base

   c. both

   d. acid

   e. None of the above

9. Is it an acid or a base if $[H^+]$ is 1×10^{-5}?

   a. base

   b. both

   c. acid

   d. neither

   e. None of the above

10. Is it an acid or a base if $[H^+]$ is 1×10^{-3}?

    a. base

    b. both

    c. acid

    d. neither

    e. None of the above

11. Is it an acid or a base if $[H^+]$ is 1×10^{-9}?

   a. neither

   b. base

   c. both

   d. acid

   e. None of the above

12. Is it an acid or a base if $[H^+]$ is 1×10^{-7}?

   a. neither

   b. base

   c. both

   d. acid

   e. None of the above

13. What color will bromothymol blue be at a pH of 6?

   a. blue

   b. yellow

   c. orange

   d. pink

   e. None of the above

14. What color will phenolphthalein be at a pH of 9?

   a. blue

   b. yellow

   c. orange

   d. pink

   e. None of the above

15. What color will methyl red be at a pH of 5?

   a. pink

   b. clear

   c. red

   d. orange

   e. None of the above

What is an Acid? A Base?

16. What color will water be at a pH of 8?

    a. pink

    b. clear

    c. red

    d. orange

    e. None of the above

17. What is the [H$^+$] if the pH is 3?

    a. 0.000001

    b. 0.0001

    c. 0.00001

    d. 0.001

    e. None of the above

18. What is the [H$^+$] if the pH is 7?

    a. 0.00001

    b. 0.00000001

    c. 0.0001

    d. 0.0000001

    e. None of the above

19. What is the [H$^+$] if the pH is 6?

    a. 0.000001

    b. 0.0000001

    c. 0.00001

    d. 0.001

    e. None of the above

20. What is the [H$^+$] if the pH is 5?

    a. 0.0000001

    b. 0.000001

    c. 0.0001

    d. 0.00001

    e. None of the above

21. You neutralize 20 mL of HCl with 100 mL of 3M NaOH. What is the molarity of this acid?

   a. 18M

   b. 10M

   c. 15M

   d. 12M

   e. None of the above

22. You neutralize 120 mL of HNO_3 with 360 mL of 4M KOH. What is the molarity of this acid?

   a. 18M

   b. 10M

   c. 15M

   d. 12M

   e. None of the above

23. You neutralize 30 mL of H_2CO_3 with 480 mL of 2M NH_4OH. What is the molarity of this acid?

   a. 18M

   b. 14M

   c. 16M

   d. 12M

   e. None of the above

24. You neutralize 30 mL of H_2SO_4 with 540 mL of 1M KOH. What is the molarity of this acid?

   a. 6M

   b. 9M

   c. 7M

   d. 8M

   e. None of the above

Chapter 13

Credit: *http://media3.washingtonpost.com/wp-dyn/content/photo/2008/09/25/PH2008092501014.jpg*

What is Chemical Equilibrium?

Reverse Reactions

We usually think about chemical reactions as going from reactants to products (left to right). This is called the **forward reaction**. But every reaction has some molecules on the products side that break up and go back into the reactants they came from. They go from products to reactants (right to left). This is called the **reverse reaction**. The two reactions happen simultaneously (at the same time) and are shown by a double pointing arrow,

$$C\,(s)\ +\ O_2\,(g)\ \leftrightarrow\ CO_2\,(g)$$

Reaching Equilibrium

Eventually the rate of the reaction going from products to reactants (reverse reaction) becomes equal to the rate of the reaction going from reactants to products (forward reaction). When that happens, we say that the reaction has reached **chemical equilibrium**. This can take from a few seconds to several days. Figure 13-1 shows a chemical reaction going to equilibrium.

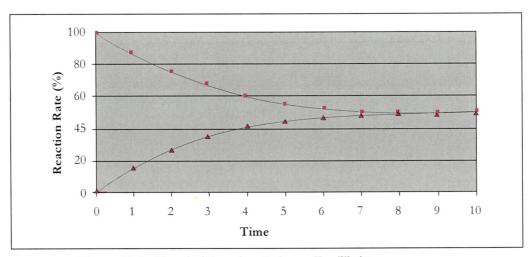

Figure 13-1: Chemical Reaction Going to Equilibrium

When the reaction begins there are no products (0%). Everything is reactants (100%). As the reaction continues, the number of product molecules starts to grow. But that uses up more and more of the reactants, so the forward reaction begins to slow down.

Meanwhile, some of the newly made products start breaking apart and going back into reactants. As more and more products are made, the number of molecules breaking up and going in the reverse direction increases. So the rate in the reverse direction increases while the rate in the forward direction decreases. Eventually the two become equal, and the number of reactants becoming products is equal to the number of products becoming reactants.

But be careful! Even though the reaction rates are equal, the numbers of molecules are not. At equilibrium, the product molecules are usually more than 99% of the total molecules, and the reactant molecules are less than 1%.

Collision Theory

Most chemists believe chemical reactions occur when molecules collide. They think a C atom and an O_2 molecule need to collide at high speeds in order to make a CO_2 molecule. This theory is called the **collision theory**.

The collision theory suggests there are several things that can affect the rate of a reaction: the energy with which molecules collide, the frequency of the collisions, the concentration of the reactants, the distribution of the reactants, and the orientation of the molecules when they collide.

Energy of Collisions

Temperature is just the average speed (kinetic energy) of the molecules. As the temperature increases, the kinetic energy of the molecules increases. Then more and more of them have enough energy to break bonds in other molecules and cause a chemical reaction. The more molecules have enough energy to break the bonds in other molecules, the faster the reaction occurs.

Frequency of Collisions

Because a higher temperature also increases the speed of the molecules, it also increases the frequency of collisions (how often they happen). The more often the molecules collide, the better the chances are that one of them will break the bonds in another molecule and cause a chemical reaction.

Concentration of Reactants

If you increase the concentration of the reactants, there are more molecules hanging around. So there are now more targets to collide with. This increases the frequency of collisions, so the chemical reaction goes faster.

Distribution of Reactants

During reactions, some areas can become saturated (filled) with product molecules. If that happens, the reactant molecules won't be able to find other reactant molecules with which to collide. But if you stir the solution, you prevent the saturation, and the reaction continues going at a fast rate.

Orientation of Reactants

Not only do the molecules have to collide; they also have to collide in just the right way. Take this reaction, for example, and look at Figure 13-2,

$$CO\ (g)\ +\ O_3\ (g)\ \leftrightarrow\ CO_2\ (g)\ +\ O_2\ (g)$$

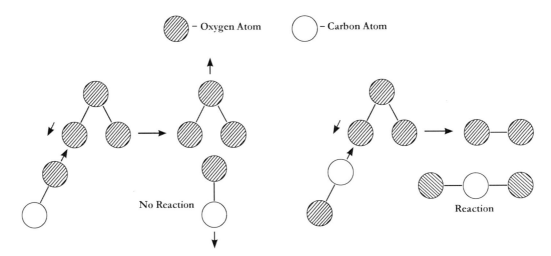

Figure 13-2: Carbon Monoxide and Ozone Collisions

If the CO collides with the O_3 in such a way that the C is touching the O_3, a reaction can occur and a molecule of CO_2 can be formed. But if the CO collides with the O_3 in such a way that two O molecules are touching each other, the C does not touch the O_3, and a molecule of CO_2 cannot be formed. The two molecules just bounce apart. So the molecules have to collide in just the right way for a reaction to occur.

That brings us to the subject of catalysts. A **catalyst** is a substance that speeds up the rate of a reaction without actually taking part in the reaction. Catalysts usually work by encouraging the molecules to align themselves in just the right way for a reaction to occur. In the catalytic converter in your car, the CO and O_2 stick to the Pt surfaces in such a way that the C and O_2 molecules can touch each other. That increases the rate of the reaction.

Activation Energy

Most reactions release (give off) heat when they occur, but some reactions need to absorb (take in) heat in order to occur. Reactions that give off heat are called **exothermic** reactions. Those that need to absorb heat are called **endothermic** reactions.

In both cases the reactants need to have a certain **activation energy** for the reaction to occur. If the reactant molecules do not have enough energy, nothing happens. The following energy diagrams show that. The energy diagram for an exothermic reaction is shown in Figure 13-3.

The energy at the top of the hill is the activation energy. The top of the hill is called the transition point. Here the molecules make the transition from one type of molecule to another. The molecules get their activation energy by colliding with other molecules that have high kinetic energy.

What is Chemical Equilibrium?

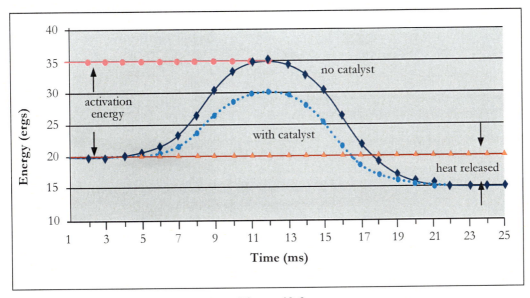

Figure 13-3:

In Figure 13-3, look at the energy level before the reaction. It is higher than the energy level after the reaction. The difference in energy is given off as heat (the test tube gets warm) during the reaction.

Figure 13-4 shows the energy diagram for an endothermic reaction.

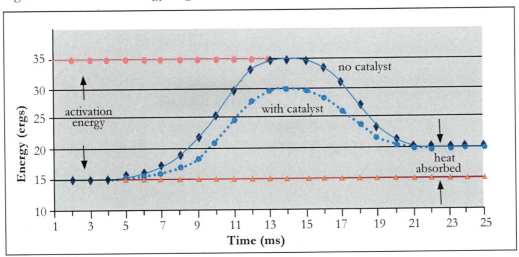

Figure 13-4 :

Again, at the top of the hill is the transition point. This time, though, the energy level before the reaction is less than the energy level after the reaction. The extra energy is obtained (gotten) by absorbing heat from the surroundings (the test tube gets cold) during the reaction.

By the way, a catalyst lowers the activation energy you need for a reaction to happen. That is shown in both diagrams by a dotted line.

Equilibrium Constants

When a chemical reaction reaches equilibrium, the products and reactants are breaking apart and recombining at the same rate. So the concentrations of the products and the reactants concentrations are no longer changing.

Look at this reaction, for example,

$$2NO\ (g) + O_2\ (g) \leftrightarrow 2NO_2\ (g)$$

After this reaction reaches equilibrium, the [NO], [O_2], and [NO_2] concentrations are no longer changing because the number of NO and O_2 molecules going into NO_2 is equal to the number of NO_2 molecules going back into NO and O_2.

Chemists like to write what they call an equilibrium constant like this one,

$$K_E = \frac{[NO_2] \times [NO_2]}{[O_2] \times [NO] \times [NO]} = \frac{[NO_2]^2}{[O_2] \times [NO]^2}$$

with the products all multiplied together (remember that $2NO_2$ means $NO_2 + NO_2$), the reactants all multiplied together, and the first one divided by the second. Notice that the [NO_2] x [NO_2] is written in shorthand as [NO_2]^2, just like you would in Algebra. Because at equilibrium the concentrations are no longer changing, they are all constants, so K_E is just one big constant.

You can write a kind of general (all purpose) equilibrium equation like this,

$$a\,A + b\,B + c\,C \leftrightarrow d\,D + e\,E + f\,F$$

where A, B, C, D, E, F stand for the formulas of the compounds in the reaction and a, b, c, d, e, f are the coefficients (numbers) needed to balance the reaction. This is what the equilibrium constant looks like for that reaction,

$$K_E = \frac{[D]^d \times [E]^e \times [F]^f}{[A]^a \times [B]^b \times [C]^c}$$

This is called the **general equilibrium constant**, K_E.

Take this reaction,

$$N_2\ (g) + 3\,H_2\ (g) \leftrightarrow 2\,NH_3\ (g)$$

Using the general K_E equation, you can write the equilibrium constant for that reaction this way,

$$K_E = \frac{[NH_3]^2}{[N_2] \times [H_2]^3}$$

You can try to change the concentrations, by adding more NH_3, for example, but when you do that, the other concentrations change such that K_E stays the same. The reason has to do with how chemical equilibrium works, which we will talk about later. For right now just know that K_E remains constant.

What is Chemical Equilibrium?

Here is an example. Suppose you measure the concentrations at STP and find [N$_2$] is 0.50M, [H$_2$] is 1.50M, and [NH$_3$] is 0.50M. What is K$_E$?

$$K_E = \frac{[0.5]^2}{[0.5] \times [1.5]^3} = 0.15 \text{ at STP}$$

You have to add "at STP" because K$_E$ will be different at a different temperature or pressure.

Heterogeneous Equilibrium Constants

In that example, all the molecules were in the same state. They were all gases. But what happens if one of them is a solid or a liquid? When you have molecules in different physical states, the equilibrium constant is called a **heterogeneous equilibrium constant**.

Well the concentration of a solid doesn't change, does it? So its concentration is constant. Instead of having two constants in the equation, the constant for the solid is included in the K$_E$ for the reaction. So the concentration of a solid does not appear in the equilibrium equation.

The same thing with a liquid. Its concentration doesn't change. So it is included in K$_E$, and its concentration doesn't appear in the equation either. Only the concentrations of gases and solutions appear in the K$_E$ equation.

As an example, write K$_E$ for this reaction at 1 atm and 100 °C,

$$C\ (s) + HOH\ (g) \leftrightarrow CO\ (g) + H_2\ (g)$$

Since C is a solid, it does not appear in the KE equation and,

$$K_E = \frac{[CO] \times [H_2]}{[HOH]} \text{ at 1 atm and 100 °C}$$

But notice that HOH does appear in this equation. That's because the (g) tells you that HOH is not a liquid. It's a gas (steam). Tricky!

Equilibrium Constant for Ionization

The K$_E$ equation can also be used for several special cases. One such case is the equilibrium constant, K$_I$, for the ionization of a weak acid or base. The equation is the same, but the products are ions instead of compounds.

Take for example the ionization of HC$_2$H$_3$O$_2$ in water at 1 atm and 22 °C,

$$HC_2H_3O_2\ (aq) \leftrightarrow H^+\ (aq) + C_2H_3O_2^-\ (aq)$$

Using the general K$_E$ equation, the K$_I$ is written this way,

$$K_I = \frac{[H^+] \times [C_2H_3O_2^-]}{[HC_2H_3O_2]} \text{ at 1 atm and 22 °C}$$

Equilibrium Constant for Solubility

Likewise, the general K_E equation can be used for the solubility constant, K_S, for an insoluble ionic compound. If you look at the solubility rules for compounds, $Mg(OH)_2$ is insoluble. But all insoluble compounds are really slightly (less than 1%) soluble. The reaction is written,

$$Mg(OH)_2 \text{ (s)} \leftrightarrow Mg^{+2} \text{ (aq)} + 2\, OH^- \text{ (aq)}$$

So you can write this equilibrium constant for its solubility at STP,

$$K_S = [Mg^{+2}] \times [OH^-]^2 \text{ at STP}$$

Le Chatelier's Principle

K_E is a constant, so it cannot be changed. If you try to change it by changing the concentration of one substance or another, the concentrations of the other substances rearrange themselves such that K_E still does not change. In fact, when any reversible reaction at equilibrium is upset by a change in concentrations, the equilibrium shifts in such a way as to reduce the change that caused it. That statement is called **Le Chatelier's principle**, and it also applies to other equilibrium constants like K_I and K_S.

Look at this example again,

$$HC_2H_3O_2 \text{ (aq)} \leftrightarrow H^+ \text{ (aq)} + C_2H_3O_2^- \text{ (aq)}$$

Say you try to change K_I by adding $NaC_2H_3O_2$. The $NaC_2H_3O_2$ will dissociate and increase $[C_2H_3O_2^-]$, but you will find that the new $C_2H_3O_2^-$ will react with H^+ to make more $HC_2H_3O_2$ such that K_I stays the same.

Now look at another example,

$$N_2 \text{ (g)} + 3\, H_2 \text{ (g)} \leftrightarrow 2\, NH_3 \text{ (g)}$$

Say you add more N_2 to try to change the K_E constant. You will find the new N_2 combines with H_2 to make more NH_3 such that K_E stays the same.

But if you add another gas that is not part of the reaction, say Ar (an inert gas), nothing happens to K_E because $[N_2]$, $[H_2]$, and $[NH_3]$ have not changed.

Pressure and Temperature Changes

Remember that a change in pressure will change the volume of a gas. That will change its concentration (molarity). Le Chatelier's principle says that all the concentrations will change in such a way as to reduce the change caused by the different pressure. In our example, if the pressure increases, more of the N_2 and H_2 will combine to make NH_3. Then there will be fewer total molecules, and that will reduce the pressure. But K_E still will be higher than it was before because, as was said before, K_E is different at different temperatures and pressures.

What is Chemical Equilibrium?

For a gas in a bottle or cylinder (a confined gas), a rise in temperature will also increase the pressure. The result will just be the same as before. More N_2 and H_2 will combine to make NH_3, and that will reduce the pressure.

For an endothermic reaction, a rise in temperature will cause the forward reaction to speed up. That will cause the concentrations of the products to increase and concentrations of the reactants to decrease. But now more of the products will start to break up. That will cause the reverse reaction to speed up, and the concentrations of products and reactants will try to return to where they were. But K_E will still be higher than it was before.

So Le Chatelier's principle also applies to pressure and temperature changes.

Homework Problems

1. According to collision theory, which of these affect the rate of a reaction?

    a. Number of reactants

    b. Atmospheric pressure

    c. Reactant distribution

    d. Product solubility

    e. None of the above

2. According to collision theory, which of these affect the rate of a reaction?

    a. Ambient temperature

    b. Degree of ionization

    c. Reactant solubility

    d. Atmospheric pressure

    e. None of the above

3. According to collision theory, which of these do not affect a reaction?

    a. Collision frequency

    b. Reactant concentration

    c. Reactant distribution

    d. Product solubility

    e. None of the above

4. According to collision theory, which of these do not affect a reaction?

    a. Reactant energy

    b. Collision ionization

    c. Reactant orientation

    d. Collision frequency

    e. None of the above

5. Write the equation for the equilibrium constant of this reaction at STP:

    KCN (aq) + HF (aq) ↔ KF (aq) + HCN (aq)

What is Chemical Equilibrium?

6. Write the equation for the equilibrium constant of this reaction at STP:

   $2\,NaOH\,(aq) + H_2SO_4\,(aq) \leftrightarrow Na_2SO_4\,(aq) + 2\,HOH\,(g)$

7. What is the equilibrium constant of this reaction if $[NH_3]$ is 0.50 M, $[HC_2H_3O_2]$ is 3.0 M, and $[NH_4C_2H_3O_2]$ is 0.50 M at STP?

   $NH_3\,(aq) + HC_2H_3O_2\,(aq) \leftrightarrow NH_4C_2H_3O_2\,(aq)$

   a. 0.33

   b. 0.15

   c. 0.67

   d. 0.30

   e. None of the above

8. What is the equilibrium constant of this reaction if $[N_2]$ is 0.50 M, $[H_2]$ is 1.50 M, and $[NH_3]$ is 0.50 M at STP?

   $N_2\,(g) + 3\,H_2\,(g) \leftrightarrow 2\,NH_3\,(g)$

   a. 0.67

   b. 0.12

   c. 0.33

   d. 0.15

   e. None of the above

9. Write the equation for the equilibrium constant of this reaction at STP:

   $2\,NaOH\,(aq) + H_2CO_3\,(aq) \leftrightarrow Na_2CO_3\,(aq) + 2\,HOH\,(l)$

10. Write the equation for the equilibrium constant of this reaction at STP:

    $Ca(OH)_2\,(aq) + H_2CO_3\,(aq) \leftrightarrow CaCO_3\,(s) + 2\,HOH\,(l)$

11. What is the equilibrium constant of this reaction if $[H_2SO_4]$ is 4.0 M, $[Ba(OH)_2]$ is 2.5 M, $[HOH]$ is 0.001 M, and $[BaSO_4]$ is 10.0 M at at STP?

    $Ba(OH)_2\,(aq) + H_2SO_4\,(aq) \leftrightarrow BaSO_4\,(s) + 2\,HOH\,(l)$

    a. 0.001

    b. 0.10

    c. 1.00

    d. 0.01

    e. None of the above

12. What is the equilibrium constant of this reaction if [HCl] is 2.0 M, [Ba(OH)$_2$] is 1.0 M, [HOH] is 0.001 M, and [Ba(Cl)$_2$] is 3.0 M at at STP?

    Ba(OH)$_2$ (aq) + 2 HCl (aq) ↔ Ba(Cl)$_2$ (aq) + 2 HOH (l)

    a. 0.50

    b. 0.67

    c. 0.75

    d. 1.5

    e. None of the above

13. Write the equation for the ionization constant of this reaction at STP:

    KOH (aq) ↔ H$^+$ (aq) + OH$^-$ (aq)

14. Write the equation for the ionization constant of this reaction at STP:

    HC$_2$H$_3$O$_2$ (aq) ↔ H$^+$ (aq) + C$_2$H$_3$O$_2^-$ (aq)

15. What is the ionization constant of this reaction if [NH$_4$OH] is 5.0 M, [NH$^+$] is 3.5 M, and [OH$^-$] is 3.5 M at STP?

    NH$_4$OH (aq) ↔ NH$_4^+$ (aq) + OH$^-$ (aq)

    a. 1.25 M

    b. 2.45 M

    c. 0.70 M

    d. 0.40 M

    e. None of the above

16. What is the ionization constant of this reaction if [Ba(OH)$_2$] is 2.0 M, [Ba$^+$] is 1.5 M, and [OH$^-$] is 1.5 M at STP?

    Ba(OH)$_2$ (aq) ↔ Ba^{+2} (aq) + 2 OH$^-$ (aq)

    a. 1.1 M

    b. 1.7 M

    c. 1.3 M

    d. 0.9 M

    e. None of the above

17. Write the equation for the solubility constant of this reaction at STP:

    PbSO$_4$ (s) ↔ Pb^{+2} (aq) + SO$_4^{-2}$ (aq)

18. Write the equation for the solubility constant of this reaction at STP:

    Al$_2$O$_3$ (s) ↔ 2 Al^{+3} (aq) + 3 O^{-2} (aq)

19. What is the solubility constant of this reaction if [ZnCO$_3$] is 5.00 M, [Zn^{+2}] is 2.50 M, and [CO$_3^{-2}$] is 2.50 M at STP?

    ZnCO$_3$ (s) ↔ Zn^{+2} (aq) + CO$_3^{-2}$ (aq)

    a. 0.80 M

    b. 2.50 M

    c. 1.25 M

    d. 2.00 M

    e. None of the above

20. What is the solubility constant of this reaction if [Al$_2$(CO$_3$)$_3$] is 2.50 M, [Al^{+3}] is 1.20 M, and [CO$_3^{-2}$] is 1.40 M at STP?

    Al$_2$(CO$_3$)$_3$ (s) ↔ 2 Al^{+3} (aq) + 3 CO$_3^{-2}$ (aq)

    a. 0.67 M

    b. 1.49 M

    c. 0.63 M

    d. 1.58 M

    e. None of the above

21. Which of the following would increase the reaction rate in the forward direction for this reaction?

    2 CO (g) + O$_2$ (g) ↔ 2 CO$_2$ (g)

    a. Increasing the temperature

    b. Increasing the pressure

    c. Adding CO to the reaction

    d. All of the above

    e. None of the above

22. Which of the following would increase the reaction rate in the forward direction for this reaction?

    NaHCO$_3$ (s) ↔ NaOH (aq) + CO$_2$ (g)

    a. Decreasing the temperature

    b. Increasing the pressure

    c. Adding CO$_2$ to the solution

    d. All of the above

    e. None of the above

23. Which of the following would increase the reaction rate in the reverse direction for this reaction?

   $N_2 (g) + O_2 (g) \leftrightarrow 2 NO (g)$

   a. Increasing the temperature
   b. Decreasing the pressure
   c. Adding more O_2
   d. All of the above
   e. None of the above

24. Which of the following would increase the reaction rate in the reverse direction for this reaction?

   $2 C (s) + O_2 (g) \leftrightarrow 2 CO (g)$

   a. Decreasing the temperature
   b. Increasing the pressure
   c. Adding more O_2
   d. All of the above
   e. None of the above

Chapter 14

Credit: http://2.bp.blogspot.com/-ivQ67gACZXg/TV1OeMnWJaI/AAAAAAAAAF8/P2hP4QFqNhc/s1600/lightning.jpg

What are Oxidation Numbers?

Introduction

Up until now, you have been balancing chemical reactions by **inspection**. You count the number of atoms on one side of the reaction and adjust the coefficients on other side to have the same number of atoms on both sides.

Well, in case you haven't discovered it already, that technique does not always work. For some reactions you have to resort to other techniques. One such technique is **oxidation and reduction**. Basically you count electrons instead of atoms.

The oxidation-reduction technique follows the transfer of electrons. It determines where the electrons came from and where they went. Like with money, if you follow the trail of electrons you can find out who did what.

Oxidation and Reduction

A substance is **oxidized** if it loses electrons to another substance in a reaction. A substance is **reduced** if it takes electrons from another one in a reaction. An oxidation-reduction reaction is also called a **redox** reaction.

An **oxidizing agent** is the substance that causes the oxidation and is reduced in the process. A **reducing agent** is a substance that causes the reduction and is oxidized in the process. These terms are very confusing. Just ignore them.

To use the redox technique, you need to find the oxidation numbers of the atoms involved. The **oxidation number** (ON) of an atom is simply the number of electrons it gains or loses in a redox reaction.

Oxidation Numbers

Oxidation numbers are a way of counting electrons. The ON of an atom is determined using the set of rules shown in Table 14-1.

To see how you apply the rules to find ONs in a compound, you need to do a few examples. First, find the ON for C in $NaHCO_3$.

By rule #4 Na is +1; by rule #2 H is +1; and by rule #3 O is -2. Since this is an ionic compound, by rule #6 the sum of the ONs of all the atoms equals 0. So C has to be equal to +4 because,

$$Na + H + \underline{C} + 3\,O = (+1) + (+1) + (X) + 3\,(-2) = 0$$

Now do another example. Find the ON for Cr in $Cr_2O_7^{-2}$.

By rule #3, O is -2. Since it is a polyatomic ion, by rule #5 the sum of the ONs is equal to -2. So Cr must be equal to +6 because,

$$2\,Cr + 7\,O = 2\,(X) + 7\,(-2) = -2$$

So what are these ONs? What are they used for? Well look at this reaction,

What are Oxidation Numbers?

$$2\,Ag\,(s) + H_2O\,(g) \rightarrow Ag_2O\,(s) + H_2\,(g)$$

By rule #4, Ag has an ON of +1; and by rule #3, O has an ON of -2. What this means is that Ag is oxidized (loses electrons) like this,

$$Ag \rightarrow Ag^+ + e^-$$

Likewise O is reduced (steals electrons) like this,

$$O + 2\,e^- \rightarrow O^{-2}$$

So the +1 oxidation number means Ag loses 1 electron, and the -2 means O gains 2 electrons. Can you see how you play this game?

Table 14-1: Rules for Assigning Oxidation Numbers

1.	A substance in its natural state has an oxidation number of 0.
2.	A hydrogen ion is usually given an oxidation number of +1.
3.	An oxygen ion is usually given an oxidation number of -2.
4.	A monoatomic ion has an oxidation number equal to its charge.
5.	In a polyatomic ion, the SUM of the oxidation numbers of each of its atoms is equal to the net ionic charge on the polyatomic ion.
6.	In an ionic compound, the SUM of the oxidation numbers of each of the atoms in the ionic compound is equal to 0.
7.	In a molecular compound, the most electronegative atom is given a negative oxidation number equal to its anion charge.

Half-Reaction Redox Method

To actually use oxidation numbers to balance a reaction, you need to break up the reaction into two half-reactions. One shows the reactant that is oxidized (loses electrons), and the other shows the reactant that is reduced. Then you adjust them to have the same number of electrons lost and stolen.

Do an example. Use the half-reaction method to balance this reaction,

$$Zn\,(s) + HNO_3\,(aq) \rightarrow ZnO\,(aq) + HNO\,(aq)$$

According to the rules, O has an ON of -2, and the O_2 from the HNO_3 gets reduced (gains electrons) like this,

$$O_2 + 4\,e^- \rightarrow 2\,O^{-2}$$

According to the rules, Zn has an ON of +2 and gets oxidized like this,

$$Zn \rightarrow Zn^{+2} + 2\,e^-$$

Because the number of electrons lost and stolen has to be the same (conservation of charge), you need to multiply the second half-reaction by 2 to get the same number of electrons that are in the first half-reaction,

$$2\ Zn \rightarrow 2\ Zn^{+2} + 4\ e^-$$

Now add the first and third half-reactions together,

$$2\ Zn + O_2 + 4\ e^- \rightarrow 2\ Zn^{+2} + 4\ e^- + 2\ O^{-2}$$

Then cancel the 4 e^- that appear on both sides, combine the Zn^{+2} and O^{-2}, and add back in the HNO we left out (because its ON had not changed). You now get this balanced equation,

$$2\ Zn + HNO_3 \rightarrow 2\ ZnO + HNO$$

You probably could have balanced that one by inspection, but try this one,

$$Fe_2O_3\ (l) + CO\ (g) \rightarrow Fe\ (l) + CO_2\ (g)$$

Not so easy, right? Try using the half-reaction method instead. If you figure out the ONs for each atom, you find C is oxidized (loses 2 electrons) in going from C^{+2} on the left to C^{+4} on the right, and Fe is reduced (steals 3 electrons) in going from Fe^{+3} on the left to Fe on the right,

$$C^{+2} \rightarrow C^{+4} + 2\ e^-$$
$$Fe^{+3} + 3\ e^- \rightarrow Fe$$

To have the same number of electrons on both sides, you need to multiply the first half-reaction by 3 and the second by 2,

$$3\ C^{+2} \rightarrow 3\ C^{+4} + 6\ e^-$$
$$2\ Fe^{+3} + 6\ e^- \rightarrow 2\ Fe$$

Now add them together,

$$2\ Fe^{+3} + 6\ e^- + 3\ C^{+2} \rightarrow 2\ Fe + 3\ C^{+4} + 6\ e^-$$

Finally, cancel the 6 e^- on both sides and bring back the oxygen atoms which we left out (because their ONs had not changed),

$$Fe_2O_3\ (l) + 3\ CO\ (g) \rightarrow 2\ Fe\ (l) + 3\ CO_2\ (g)$$

In aqueous solutions, you can borrow H^+ and OH^- ions from the solution (as needed) to balance O and H atoms in a redox reaction. As an example, this is what happens when you put Cu into a HNO_3 solution,

$$Cu\ (s) + NO_3^-\ (aq) \rightarrow Cu^{+2}\ (aq) + NO\ (g)$$

The extra O_2 from the NO_3^- gets made into HOH (water) using H^+ ions from the solution. You have these two half-reactions,

$$Cu \rightarrow Cu^{+2} + 2\ e^-$$

$$4\ H^+ + NO_3^- + 3\ e^- \rightarrow NO + 2\ HOH$$

To have the same number of electrons on both sides, you need to multiply the first half-reaction by 3 and the second one by 2,

What are Oxidation Numbers?

$$3 \text{ Cu} \rightarrow 3 \text{ Cu}^{+2} + 6 \text{ e}^-$$

$$8 \text{ H}^+ + 2 \text{ NO}_3^- + 6 \text{ e}^- \rightarrow 2 \text{ NO} + 4 \text{ HOH}$$

Combining the two half-reactions, you get this balanced ionic equation,

$$3 \text{ Cu (s)} + 8 \text{ H}^+ + 2 \text{ NO}_3^- \text{ (aq)} \rightarrow 3 \text{ Cu}^{+2}\text{(aq)} + 2 \text{ NO (g)} + 4 \text{ HOH}$$

Reduction Potential

Some redox reactions happen **spontaneously** (all by themselves). These reactions allow you to generate the electricity you get from batteries. They also cause some metals to corrode when welded together.

To understand what reactions occur spontaneously, you need to understand the reduction potential order in metals. The **reduction potential** of metals is nothing more than the activity series you learned about earlier. Both are really just the relative electronegativities of the metals.

Just like with the activity series, there is a certain order of the metals. If a metal is higher on the list than another, it will spontaneously reduce it. If it is lower, nothing happens. Here is the order,

Table 14-2: Decreasing Reduction Potential

Li>K>Ba>Sr>Ca>Na>Mg>Al>Mn>Zn>Cr>Fe>Cd>Co>Ni>Sn>Pb>H>Cu>I>Ag>Hg>Br>Cl>F

Since oxidation is a loss of electrons and reduction is a gain of electrons, the **oxidation potential** goes in the reverse direction to reduction potential.

Electrochemistry

A Li-Ni battery is what powers your laptop. When you use it, this happens,

$$\text{Li} \rightarrow \text{Li}^{+2} + 2 \text{ e}^-$$

During this step, electricity passes through your laptop and you can surf the internet. And then this spontaneous reaction occurs,

$$\text{Li}^{+2} + \text{Ni} \rightarrow \text{Li} + \text{Ni}^{+2}$$

Because Li has a higher reduction potential than Ni, it can steal 2 electrons from Ni. Now you can repeat the process again and surf the net some more.

But if you try it the other way around,

$$\text{Li} + \text{Ni}^{+2} \rightarrow \text{NR}$$

because Ni has a lower reduction potential than Li. So eventually all the Ni becomes Ni^{+2}, and your battery wears out. But then you recharge it like this,

$$\text{Ni}^{+2} + 2 \text{ e}^- \rightarrow \text{Ni}$$

and you start all over again. This recharge reaction is not spontaneous because you have to add electricity (plug it into the power supply).

Batteries like this that work with the two metals in contact with each other are called **electrolytic cells**.

Another type of **electrochemical cell** is the battery in your car. Here a H_2SO_4 solution is between the two metals. This is the overall reaction between the Zn (+) and Cu (-) terminals of your battery,

$$Zn\ (s) + CuSO_4\ (aq) \rightarrow Cu\ (s) + ZnSO_4\ (aq)$$

At the positive terminal, this happens,

$$Zn\ (s) \rightarrow Zn^{+2}\ (aq) + 2\ e^-$$

And at the negative terminal, this happens,

$$Cu^{+2}\ (aq) + 2\ e^- \rightarrow Cu\ (s)$$

When your car battery runs down, you recharge it by running current through it in the opposite direction (using the alternator in your car) and reversing the reaction, just like you did with the Li-Ni battery.

Batteries like this with an acid solution between the two metals are called **voltaic cells**. The use of electricity from spontaneous redox reactions is called **electrochemistry**.

New Chemist

Well you are now a chemist! Okay, maybe just a beginner chemist.

What are Oxidation Numbers?

Homework Problems

1. Using the rules for determining oxidation numbers, what is the ON for S in SO_2?
   a. 8
   b. 6
   c. 4
   d. 2
   e. None of the above

2. Using the rules for determining oxidation numbers, what is the ON for N in NO_2?
   a. 5
   b. 4
   c. 3
   d. 2
   e. None of the above

3. Using the rules for determining oxidation numbers, what is the ON for N in HNO_3?
   a. 4
   b. 5
   c. 2
   d. 3
   e. None of the above

4. Using the rules for determining oxidation numbers, what is the ON for S in H_2SO_4?
   a. 8
   b. 4
   c. 6
   d. 2
   e. None of the above

5. Using the rules for determining oxidation numbers, what is the ON for N in NH_3?
   a. -4
   b. -5
   c. -2
   d. -3
   e. None of the above

6. Using the rules for determining oxidation numbers, what is the ON for C in CH_4?
   a. -6
   b. -4
   c. -5
   d. -3
   e. None of the above

7. Using the rules for determining oxidation numbers, what is the ON for Mn in MnO_4?
   a. 1
   b. 5
   c. 3
   d. 7
   e. None of the above

8. Using the rules for determining oxidation numbers, what is the ON for As in AsO_4^{-3}?
   a. 7
   b. 1
   c. 5
   d. 3
   e. None of the above

9. Using the rules for determining oxidation numbers, what is the ON for C in $C_2O_4^{-2}$?
   a. 6
   b. 4
   c. 5
   d. 3
   e. None of the above

10. Using the rules for determining oxidation numbers, what is the ON for P in PO_4^{-3}?
    a. 1
    b. 5
    c. 3
    d. 7
    e. None of the above

11. Using the rules for determining oxidation numbers, what is the ON for Cl in $NaClO_3$?
    a. 3
    b. 5
    c. 6
    d. 4
    e. None of the above

What are Oxidation Numbers?

12. Using the rules for determining oxidation numbers, what is the ON for Cr in $K_2Cr_2O_7$?

   a. 5
   b. 3
   c. 4
   d. 6
   e. None of the above

13. Use the half-reaction redox method to balance the following reaction:
    Fe_2O_3 (s) + CO (g) → Fe (s) + CO_2 (g)

14. Use the half-reaction redox method to balance the following reaction:
    Al_2O_3 (s) + CO (g) → Al (s) + CO_2 (g)

15. Use the half-reaction redox method to balance the following reaction:
    Pb (s) + HNO_3 (aq) → PbO_2 (s) + NO_2 (g) + HOH (l)

16. Use the half-reaction redox method to balance the following reaction:
    S (s) + HNO_3 (aq) → H_2SO_4 (aq) + NO_2 (g) + HOH (l)

17. Use the half-reaction redox method to balance the following reaction:
    Cu (s) + $AgNO_3$ (aq) → $Cu(NO_3)_2$ (aq) + Ag (s)

18. Use the half-reaction redox method to balance the following reaction:
    Zn (s) + $AgNO_3$ (aq) → $Zn(NO_3)_2$ (aq) + Ag (s)

19. Use the half-reaction redox method to balance the following reaction:
    Zn (s) + HNO_3 (aq) → ZnO (aq) + NO (g) + HOH (l)

20. Use the half-reaction redox method to balance the following reaction:

   Ni (s) + HNO$_3$ (aq) → NiO (aq) + NO (g) + HOH (l)

21. Use the half-reaction redox method to balance the following reaction:

   Al (s) + MnO$_2$ (s) → Mn (s) + Al$_2$O$_3$ (s)

22. Use the half-reaction redox method to balance the following reaction:

   CH$_4$ (g) + O$_2$ (g) → CO$_2$ (g) + HOH (g)

23. Use the half-reaction redox method to balance the following reaction:

   MnO$_2$ (s) + HBr (aq) → MnBr$_2$ (aq) + Br$_2$ (l) + HOH (l)

24. Use the half-reaction redox method to balance the following reaction:

   H$_2$S (g) + HNO$_3$ (aq) → S (s) + NO (g) + HOH (l)

Bibliography:

1. Bauer, Richard, James Birk & Pamela Marks, <u>Introduction to Chemistry</u>, McGraw Hill, Boston, MA, 2007.

2. Corwin, Charles, <u>Introductory Chemistry</u>, Pearson Prentice Hall, Upper Saddle River, NJ, 2008.

3. Cracolice, Mark, <u>Introductory Chemistry</u>, Thomson Brooks/Cole, Belmont, CA, 2007.

4. Glasstone, Samuel & David Lewis, <u>Elements of Physical Chemistry</u>, Van Nostrand, Princeton, NJ, 1960.

5. Goldberg, David, <u>Fundamentals of Chemistry</u>, McGraw Hill, Dubuque, IA, 2007.

6. Hein, Morris & Susan Arena, <u>Foundations of College Chemistry</u>, Wiley and Sons, Hoboken, NJ, 2007.

7. Malone, Leo, <u>Basic Concepts of Chemistry</u>, Wiley and Sons, New York, NY, 2004.

8. Russo, Steve & Michael Silver, <u>Introductory Chemistry</u>, Benjamin Cummings, San Francisco, CA, 2007.

9. Stoker, Stephen, <u>Introduction to Chemical Principles</u>, Pearson Prentice Hall, Upper Saddle River, NJ, 2008.

10. Timberlake, Karen & William Timberlake, <u>Basic Chemistry</u>, Benjamin Cummings, San Francisco, CA, 2008.

11. Tro, Nivaldo, <u>Introductory Chemistry</u>, Pearson Education, Upper Saddle River, NJ, 2006.

12. Zumdahl, Steven, <u>Introductory Chemistry</u>, Houghton Mifflin, Boston, MA, 2006.

Ca^{+2}	Au^+	Ag^+
Cs^+	Ba^{+2}	Al^{+3}
Cl^-	Br^-	Ar
Cr^{+3}	C^{-4}	As^{-3}

silver	gold	calcium
aluminum	barium	cesium
argon	bromine	chlorine
arsenic	carbon	chromium

Li$^+$	He	F$^-$
Mg^{+2}	Hg$^+$ or Hg^{+2}	Fe^{+2} or Fe^{+3}
Mn^{+2}	I$^-$	Ge^{+4}
N^{-3}	K$^+$	H$^+$

fluorine	helium	lithium
iron (II) or iron (III)	mercury (I) or mercury (II)	magnesium
germanium	iodine	manganese
hydrogen	potassium	nitrogen

Si^{+4}	P^{-3}	Na^+
Sn^{+2} or Sn^{+4}	Pb^{+2} or Pb^{+4}	Ne
W^{+3}	S^{-2}	Ni^{+2}
Zn^{+2}	Sb^{-3}	O^{-2}

sodium	phosphorus	silicon
neon	lead (II) or lead (IV)	tin (II) or tin (IV)
nickel	sulfur	tungsten
oxygen	antimony	zinc

Cu^+ or Cu^{+2}	Co^{+2} or Co^{+3}	Cd^{+2}
ClO_3^-	ClO_2^-	ClO^-
CNO^-	CN^-	ClO_4^-
$C_2H_3O_2^-$	CO_3^{-2}	CO_2^{-2}

cadmium	cobalt (II) or cobalt (III)	copper (I) or copper (II)
hypo-chlorite	chlorite	chlorate
per-chlorate	cyanide	cyanate
carbonite	carbonate	acetate

HCO_3^-	$Cr_2O_7^{-2}$	CrO_4^{-2}
MnO_4^-	MnO_3^-	HSO_4^-
NO_3^-	NO_2^-	NH_4^+
SO_4^{-2}	PO_4^{-3}	OH^-

chromate	di-chromate	bi-carbonate
bi-sulfate	manganate	per-manganate
ammonium	nitrite	nitrate
hydroxide	phosphate	sulfate